Cassava

IADS DEVELOPMENT-ORIENTED LITERATURE SERIES

Steven A. Breth, series editor

Cassava: New Potential for a Neglected Crop
was prepared under the auspices of the
International Agricultural Development Service
and
CIAT
(Centro Internacional de Agricultura Tropical)
with the partial support of the
German Agency for Technical Cooperation
(Deutsche Gesellschaft für Technische Zusammenarbeit)

ALSO IN THIS SERIES

Rice in the Tropics: A Guide to the Development of National Programs, Robert F. Chandler, Jr.

Small Farm Development: Understanding and Improving Farming Systems in the Humid Tropics, Richard R. Harwood

Successful Seed Programs: A Planning and Management Guide, Johnson E. Douglas

Tomatoes in the Tropics, Ruben L. Villareal

Wheat in the Third World, Haldore Hanson, Norman E. Borlaug, and R. Glenn Anderson

Cassava
New Potential for a Neglected Crop

James H. Cock

Routledge
Taylor & Francis Group
LONDON AND NEW YORK

First published 1985 by Westview Press, Inc.

Published 2018 by Routledge
52 Vanderbilt Avenue, New York, NY 10017
2 Park Square, Milton Park, Abingdon, Oxon OX14 4RN

Routledge is an imprint of the Taylor & Francis Group, an informa business

Library of Congress Cataloging in Publication Data
Cock, James H.
Cassava, new potential for a neglected crop.
(IADS development-oriented literature series)
"Prepared under the auspices of the International Agricultural Development Service and CIAT . . . with the support of GTZ"—
Bibliography: p.
Includes index.
1. Cassava. I. International Agricultural Development Service. II. Centro Internacional de Agricultura Tropical. III. Deutsche Gesellschaft für Technische Zusammenarbeit. IV. Title. V. Series.
SB211.C3C56 1985 633.6′82 84-15374
ISBN 13: 978-0-367-01920-4 (hbk)

Contents

Tables

Figures

Foreword

The International Agricultural Development Service is pleased to add this book to its series of development-oriented literature. It is the first comprehensive work on cassava, one of the world's most important food crops.

Until quite recently, cassava was hardly known outside of the tropics. Since it was not a significant export crop, nor was it grown in developed countries, it attracted little attention from scientists before the founding of CIAT (Centro Internacional de Agricultura Tropical) in Colombia and IITA (International Institute of Tropical Agriculture) in Nigeria in the 1960s. These two international agricultural research centers have done extensive experimentation with cassava and have stimulated cassava research throughout the world. Consequently, knowledge about the crop has expanded at an impressive rate.

But in fast-moving fields, researchers rarely pause to synthesize and sum up the new knowledge. In the foreword to one of the first books in the IADS development-oriented literature series, the late Sterling Wortman wrote:

> Most agricultural literature . . . is in fragments, scattered over a wide range of journals and other publications. One cannot obtain a comprehensive view without resorting to the study of many narrow articles in numerous publications. . . . There are scores of commodities and problem areas for which comprehensive presentations of available information are needed, written in language understandable to both scientists and nonscientists. [Robert F. Chandler, Jr., *Rice in the Tropics* (Boulder, Colo.: Westview Press, 1979), pp. xv–xvi]

This book explores a vital agricultural topic and well exemplifies the kind of exposition that Sterling Wortman envisioned: practical, authoritative, up-to-date. IADS is grateful to CIAT for giving the author, James Cock, time to write this book. Dr. Cock is a plant physiologist who began his research career with work on dwarf wheats and rice. He founded CIAT's cassava program and has gained prominence as one of the world's leading cassava researchers.

IADS expresses its appreciation to GTZ (German Agency for Technical Cooperation) for providing partial funding for the development of the book.

A. Colin McClung
President, International Agricultural
Development Service

Preface

Growing food deficits are of great concern to all who are interested in the welfare of millions of people living in poverty in the third world. Attempts to resolve the problems of food production have placed great emphasis on increasing the production and productivity of grain crops, but little attention has been given to crops such as cassava, which are produced only in the tropics.

Among crops produced in the tropics, cassava is the fourth most important source of calories for humans. It is particularly noteworthy as a basic energy source for food, animal feed, and industrial uses that can be produced on marginal agricultural lands.

Information on cassava is scattered through an enormous array of publications, reports, and other documents published in numerous languages, making it difficult for agricultural officials and government planners to rapidly obtain a clear picture of the crop's potential and problems. It was for this reason that I was invited by the International Agricultural Development Service (IADS) to write this book, which describes the crop and its uses, the technology available for its production and processing, and the essential features of successful development programs.

Chapter 1 describes the crop and quantifies its importance. Chapter 2 sets forth the uses of cassava and the various methods of processing the crop after harvest. Chapters 3 and 4 are devoted to present production practices and the components of technology that can be used to increase productivity. Chapter 5 summarizes recent developments in post-harvest handling. Chapter 6 analyzes several cases in which cassava production has been increased and sets out a framework for the successful implementation of cassava development projects. Chapter 7 describes the critical features of national cassava programs with special reference to research and government policy issues.

I am grateful to the IADS for inviting me to write this book and to Steven Breth of IADS for assisting in all stages of the production of this book and editing the final version.

The Centro Internacional de Agricultura Tropical (CIAT) provided me with a sabbatical year during which most of the writing was completed. In addition, members of the CIAT cassava program staff provided me with information and assistance in the development of the book. I am particularly indebted to John Lynam, who provided much of the information and helped in the drafting of Chapter 7.

The original manuscript was reviewed by Barry Nestel, private consultant; W. O. Jones of the Food Research Institute, Stanford University; and Bede Okigbo of the International Institute of Tropical Agriculture. The valuable comments of the reviewers, particularly with respect to cassava in Africa, were incorporated into the later drafts.

James H. Cock
Cali, Colombia

Introduction

Among tropical crops, cassava, rice, sugarcane, and maize are the most important suppliers of calories. Until recently, however, decision makers concerned with agriculture in developing countries overlooked cassava. Owing to widely held misconceptions about the crop, only small amounts of funds have been allocated to cassava research and development. This book attempts to demonstrate how cassava, through sustained research and improved production methods, can contribute to improving the diet and livelihood of millions of people. Cassava is grown mainly by small farmers with labor-intensive methods. Although consumption is highest in rural areas, cassava should by no means be regarded as purely a subsistence crop—most cassava is sold or traded outside the farm where it was produced. In the lowland tropics, cassava, particularly in dried forms, is often the cheapest source of calories, and new growing methods promise to reduce the cost even further.

Cassava is sometimes disparaged as an undesirable food that contains little besides carbohydrate. But in many developing countries, calories are in fact the paramount nutritional shortage. The cyanide present in raw cassava is a drawback, but among millions of cassava consumers worldwide, chronic cyanide toxicity occurs only in certain areas of Africa. In those areas, it appears to be associated with severely deficient diets combined with an underprocessing of cassava. The remedy certainly does not lie in condemning cassava, but rather, in improving overall nutrition and in more adequate processing.

Marketing is a major problem with a perishable crop like cassava. The demand for cassava as food in a locality may be quickly saturated if production increases sharply and no alternative use exists. The development of small enterprises to produce dried cassava for use in animal feed can help stabilize the price of cassava as well as create employment and save foreign exchange.

Although the merits of cassava are becoming recognized, it is often feared that expanding cassava production will degrade soil fertility, particularly on marginal soils. Cassava depletes soils no more than other crops. It does, however, have the ability to grow on already depleted soils, which is one of its great advantages. This characteristic, combined with its high production potential, points to the bright prospect that cassava offers as a basic energy source from marginal land in the tropics.

1
Cassava: The plant and its importance

Cassava is a starchy root crop that is grown almost entirely within the tropics. Although it is one of the most important food crops in tropical countries (see Table 1), it is little known elsewhere, and within the tropics, it is often considered to be a low-grade subsistence crop.

World production

In the 1979–1981 period, world production of cassava was about 120 million tons annually. Since the cassava root is about 65 percent water, that production level yields about 42 million tons of dry matter, which is equivalent, in energy terms, to 40 to 50 million tons of grain. Because much of the cassava grown is produced by small farmers in marginal agricultural areas and because most of it does not enter commercial marketing channels, it is difficult to obtain data on production. Food and Agriculture Organization (FAO) statistics are the best available guide to global production; however errors in the estimates for individual countries can be quite large. FAO figures for Brazilian production, for example, are much greater than estimates obtained from data in Brazil's agricultural census. From 1960 to 1980, world output of cassava rose at about the same rate as population. Almost all the gain has been due to expansion in planted area; yields have changed little.

Africa, though it has slightly more than half the world's cassava land, produces only 37 percent of the world crop. FAO estimates

Table 1
Staple crops as sources of calories in human diets in the tropics and worldwide (in billion kcal/day)

Crop	Tropics	World
Rice	924	2043
Sugar (cane and beet)	311	926
Maize	307	600
Cassava	172	178
Sorghum	147	208
Millet	128	204
Wheat	< 100[a]	1877
Potato	54	434
Banana	32	44
Plantain	30	30
Sweet potato	30	208

Source: FAO, *Food balance sheets 1975–1977 average* (Rome, 1980).

[a] Excluding Brazil, Mexico, and India, as the major wheat production zones of those countries are outside the tropics.

show that yields have been nearly constant at 6.4 t/ha in recent years, but between 1969–1971 and 1980, plantings increased from 5.8 million hectares to 7.3 million hectares, and annual production climbed from 39.4 million tons to 46.5 million tons. Zaire, Nigeria, and Tanzania are the major African cassava producers. Average yields in those countries range from 9 t/ha in Nigeria to less than 5 t/ha in Tanzania (see Table 2).

Asia accounts for only 27 percent of the world cassava area, but it produces 37 percent of the crop. Yields vary from less than 8 t/ha in Vietnam to over 16 t/ha in India. Increasing Asian output during the last 20 years has been due mainly to expansion in Thailand. Production in Thailand grew from less than 1 million tons in 1957 to 4 million tons in 1972 to an estimated 17.9 million tons in 1981. The second largest producer in Asia is Indonesia, which harvests over 13 million tons annually. Its land area in cassava had declined slightly in recent years, but gains in yields, which now average about 9.6 t/ha, have more than compensated for that decline. Another important

Table 2
Cassava production in Africa, Asia, and the Americas and selected countries, 1979–1981 average

	Area (million ha)	Yield (t/ha)	Production (million tons)
Africa	7.32	6.4	46.44
Zaire	1.82	6.8	12.40
Nigeria	1.18	9.2	10.83
Tanzania	0.94	4.9	4.60
Mozambique	0.60	4.6	2.78
Angola	0.13	14.2	1.85
Ghana	0.23	8.0	1.80
Madagascar	0.28	6.0	1.67
Uganda	0.41	3.4	1.39
Burundi	0.08	15.4	1.18
Cameroon	0.23	4.3	1.00
Central African Republic	0.33	3.0	0.99
Ivory Coast	0.22	3.5	0.75
Kenya	0.08	7.9	0.64
Congo	0.08	6.6	0.53
Togo	0.03	16.2	0.45
Asia	3.80	11.6	43.97
Thailand	1.00	14.4	14.54
Indonesia	1.42	9.6	13.67
India	0.35	16.7	5.90
Vietnam	0.46	7.3	3.38
China	0.24	13.0	3.06
Philippines	0.20	11.5	2.28
Sri Lanka	0.06	9.5	0.52
Malaysia	0.04	10.1	0.37
South America	2.55	11.7	29.90
Brazil	2.07	11.8	24.47
Colombia	0.21	9.8	2.07
Paraguay	0.13	14.9	1.97
Peru	0.03	11.4	0.40
Venezuela	0.04	9.5	0.36
Bolivia	0.02	12.4	0.22
Central America and the Caribbean	0.15	6.0	0.90

Source: FAO, *Production yearbook* (Rome, 1981).

Asian producer is India. Production has expanded markedly over the last few years in southwestern India, the primary growing area, due to increases in yields, which now are 16 to 17 t/ha. These high yields have been achieved despite soils that are very acid and infertile.

In the Western Hemisphere, Brazil dominates the production of cassava, accounting for four-fifths of the cassava in Latin America and one-fifth of world production. Cassava is grown throughout Brazil—in the hot, humid jungle of the Amazon basin, in the dry areas of the Northeast, and in the cooler southern areas including part of Rio Grande do Sul where there are frosts and even occasional snow. Brazilians grow cassava for food, animal feed, and industrial uses. In the 1970s, yields declined from 14 t/ha to less than 12 t/ha, and total production decreased slightly to 25 million tons a year. The drop in yield may be related to the enormous expansion of soybean production and the movement of coffee production to areas further north, since these two high-value crops tend to displace cassava from more fertile soils.

Other large Latin American producers are Colombia, with 2.1 million tons, and Paraguay, with 2.0 million tons. Yields in Colombia are 10 t/ha; those in Paraguay are 15 t/ha. Paraguay, like Brazil, has high yields in part because of the 18-month growth cycle used by many producers. Although Brazil, Colombia, and Paraguay produce more than 90 percent of the cassava in the Americas, the crop is of great importance to several smaller countries as well.

In Oceania, the total cassava area is about 20,000 hectares. Papua New Guinea and Fiji are the most important producers.

Consumption patterns

Cassava is used primarily as a food, and because of its bulk and perishability, it is usually consumed in or near the place where it is grown. The only global data on cassava consumption are in the FAO *Food balance sheets,* which are at best a rough guide to the real situation. They indicate that in the mid-1970s, 65 percent of the total production was used as a basic source

Small village factories process about half the cassava that people eat. (*Source:* CIAT)

of energy in human diets (see Table 3). About equal amounts were consumed as cooked fresh cassava or as processed cassava (mainly flours and meals).

The main nonfood uses of cassava are animal feed and starch. About one-fifth of world production is fed to animals, mainly in producing countries, and the amount is rising. Almost all exports are pelleted animal feed destined for Europe. About 6 percent of world production goes into starch for industrial processes and food processing.

In the FAO data only 10 percent of total production is classified as being lost as waste. But the FAO data apparently allow for considerable losses in cassava processing, because the extraction rates for cassava products from fresh cassava are much lower than those theoretically obtainable. Hence the real waste figure is probably much higher.

Table 3
World utilization of cassava, 1975–1977 (data are presented as a percentage of total production)

Area	Human food		Animal feed[a]	Industrial use and starch	Export[b]	Waste	Change in stocks
	Fresh	Processed					
World	30.8	33.8	11.5	5.5	7.0	10.0	1.4
Africa	37.9	50.8	1.4	c	c	9.5	c
Americas	18.5	23.9	33.4	9.6	c	14.0	c
Asia	33.6	21.7	2.9	8.6	23.0	6.3	3.9
Asia without Thailand	45.7	27.9	3.9	11.7	2.3	8.6	c

Source: FAO, *Food balance sheets 1975–1977 average* (Rome, 1980); modified based on additional information.

[a] Excludes chips and pellets exported as animal food.
[b] Includes exports for animal feed.
[c] Less than 1 percent.

Consumption patterns differ tremendously among regions. In Africa, nearly all cassava is used for human consumption. Although cassava is considered to be a subsistence crop in Africa, substantial quantities are traded. Throughout Africa, cassava and cassava products can be found in the rural and urban markets. In Nigeria, *gari,* a fermented cassava flour, is transported up to 700 kilometers to urban markets, and each year the equivalent of 120,000 tons of fresh cassava enters the Lagos market as *gari.* A study in Ghana in the 1950s showed that 22,000 tons of fresh cassava and 27,000 tons of cassava products were entering Accra annually, providing 41 percent of the calories brought into the city. In Zaire, an estimated 55 percent of total production is sold off the farm. Thus, while it is impossible to accurately estimate how much cassava is traded and how much is consumed by the growers themselves, it is certain that a significant proportion of African production enters the market economy.

In Asia, over half the cassava is produced for direct human consumption, with much of the remainder exported as chips and pellets. There is considerable variation between countries. Cassava in India and Indonesia is almost entirely used for human consumption; in Malaysia, it is used for starch and the local feed industry; in Thailand, it is almost exclusively used for export. Despite these differences, throughout Asia cassava tends to be grown on small farms, which sell a large proportion of their production. Even in India, where per capita consumption is high and farm size very small, 35 to 40 percent of the cassava is sold off the farm.

In the Americas, over 40 percent of the cassava production is destined for human food. More than half of the cassava eaten is in processed forms such as flour, meals, and starch. A third of the production is used as animal feed within the producer countries. In contrast to Asia, where most of the cassava is dried before being used as animal feed, considerable quantities of fresh cassava roots are fed to animals in the Americas, particularly in Brazil and Paraguay. Little information exists on the amounts traded and used as a subsistence food in the Americas. A survey in Colombia showed that more than 90 percent of total production was sold off the farm. In Brazil, the quantity of cassava processed to either *farinha* or flour suggests that a high proportion of cassava production is utilized off the producer farms.

Importance in diets

The FAO *Food balance sheets* give estimates of cassava consumption for whole countries, but consumption patterns may vary substantially from area to area within a country. In Brazil, for example, cassava has far more dietary significance in the Northeast than it does in the South. Similarly, in India the national average figures suggest that cassava is unimportant, but in fact, in southern India cassava provides more than 700 cal/day for 20 to 30 million people. Thus national averages tend to overestimate the number of people who consume cassava as a major staple while underestimating the amounts consumed by actual cassava eaters. That notwithstanding, the food balance sheet data indicate that in 1975–1977, 500 million people in 24 countries consumed more than 100 cal/day in the form of cassava, with an average intake of over 300 calories (see Table 4).

In the countries of tropical Africa, cassava provides an average of 230 calories per person per day. In Zaire and Congo, however, the average intake is over 1000 cal/day, or close to 1 kilogram of fresh cassava per day. That amount provides about half of the total energy intake for the people of those two countries. There are an estimated 70 million people who obtain more than 500 cal/day from cassava.

The main value of cassava is the starchy roots, but the leaves are also eaten in Africa. Cassava leaves contain about 7 percent protein on a fresh weight basis (20 to 30 percent protein on a dry weight basis). Cassava leaves compare favorably with soybeans in protein quality and are considerably higher in lysine. However, methionine and possibly tryptophan are deficient. The leaves are used as a base in making sauces and soups, particularly in the Congo basin, Tanzania, and parts of West Africa. Consumption levels have been estimated as high as 500 g/day in Zaire, 40 to 170 g/day in Congo, and 30 to 100 g/day in Cameroon. The consumption of cassava leaves at these levels can make a significant contribution to total protein intake.

Table 4
Consumption of cassava in selected countries or regions

Country	Cassava consumption: Amount per person (cal/day)	Cassava consumption: As a source of dietary energy (%)	Population 1980 (millions)
Zaire	1287	56	29.0
Congo	1128	55	1.6
Central African Republic	839	39	2.2
Kerala (India)	744*	30	ca. 25.0
Mozambique	697	36	10.0
Angola	660	32	6.7
Tanzania	503	24	19.0
Liberia	501	21	1.9
Gabon	439	18	0.6
Togo	402	20	2.5
Ghana	380	19	12.0
Paraguay	351	12	3.3
Burundi	316	14	4.5
Madagascar	312	13	8.7
Fiji	300	12	0.6
Nigeria	283	13	77.0
Ivory Coast	245	10	8.0
Rwanda	238	10	5.1
Cameroon	209	9	8.5
Brazil	203	8	122.0
Indonesia	202*	10	144.0
Guinea	201	10	5.0
Kenya	123	6	16.0
Colombia	114	5	27.0

Sources: Cassava consumption: FAO, *Food balance sheets 1975–1977 average* (Rome, 1980), except for data marked with an asterisk, which are from other sources; Population: International Agricultural Development Service, *Agricultural development indicators* (New York, 1981).

In Asia, cassava is a dietary staple only in Indonesia and southern India, though it is widely grown, especially in Southeast Asia. In Indonesia, the average daily consumption is close to 200 calories per person, and in some cases more than half the total calorie intake is from cassava. In Kerala, India, it provides an estimated 744 calories per person per day for 20 to 30 million people. Although Thailand is the region's largest producer, Thais eat little cassava. Most of the crop is exported.

In Latin America, cassava is a countrywide staple only in Paraguay and Brazil. An estimated 125 million people in these countries derive more than 200 cal/day from cassava. In Brazil, consumption is high in the rural Northeast (480 cal/day) and very low in the southern cities (fewer than 20 cal/day).

Cassava is also a staple food in the jungle regions of Bolivia, Peru, and Ecuador, on the north coast and in the Santander Department of Colombia, and in the rural areas of many Caribbean islands. Throughout the cassava growing areas of South America, cassava is an important staple. Average daily consumption in the tropical countries of South America is 150 to 160 calories per person. In Central America, cassava is of little nutritional importance.

The cassava plant

Cassava (*Manihot esculenta* [Crantz]) belongs to the family Euphorbiaceae, which includes rubber (*Hevea brasilensis*) and castor bean (*Ricinus communis*). *Manihot glaziovii* (Ceara rubber) is a minor source of rubber. Before the discovery of the New World, the genus *Manihot* grew only in the Americas between about 30°N and 30°S latitude. There are two main centers of diversity—a major one in Brazil and a secondary one in Central America. Cassava is grown in most areas where *Manihot* species exist, however it is not found in the wild state. The wild progenitor of cassava is not known, nor have the areas in which cassava was domesticated been determined with certainty.

Botanical description

The cassava plant is a perennial woody shrub. Commercially, it is grown by planting a cutting taken from the woody part of the stem. After the cutting is planted, one or more of the axillary buds sprout, and roots grow principally from the base of the cutting. The shoots show strong apical dominance, which suppresses the development of side shoots. When the main shoot

Cassava is grown for its edible root. It is propagated by planting stem cuttings. (*Source:* CIAT)

becomes reproductive and begins to flower, the apical dominance is broken and several (two to four) of the axillary buds immediately below the apex begin to develop, giving the typical branching habit found in the plant. The time at which branching occurs is extremely variable; some clones never branch.

During the first months, the cassava plant mainly develops shoots and a fibrous root system. Early in the growth cycle, however, the plants store small amounts of starch in their roots. About 3 months after planting, some of the fibrous roots begin to expand rapidly, storing large quantities of starch. The thickened storage roots have a fleshy center, which composes about 85 percent of their total weight, and a tougher, more fibrous peel.

The propagation of cassava is unusual for several reasons. First, the economically useful part, the root, is not used for propagation. To propagate other major food staples, such as grain crops or potatoes, part of the harvest must be kept for replanting. Second, the multiplication rate of cassava is low—from 3 to 30 times per crop cycle, depending on conditions. Third, although cassava is normally propagated through stem cuttings, which produce genetically identical progeny, or clones, it can also be reproduced from true seeds. The progeny obtained from true seeds have great genetic variability, permitting superior varieties to be selected, which can then be reproduced by stem cuttings and hence have their new characters fixed in a stable form.

Origin and dispersal

It was once thought that cassava was originally domesticated in Brazil, where the largest number of *Manihot* species exists. Moreover, the diversity within the species is greater there than elsewhere. But little archaeological evidence exists to confirm a Brazilian domestication of cassava. Cassava is grown in areas where the rainy season is longer than the dry season, and under these conditions woody materials are rarely well conserved. Furthermore, the roots are highly perishable, so little direct evidence exists of the early cultivation of cassava (unlike the grain crops, which are often grown in arid areas and are normally dried and stored before consumption so that considerable quantities have been found in archaeological sites). In Mexico, remains of cassava leaves have been found that are 2500 years old, and cassava starch has been identified in human coprolites that are 2100 to 2800 years old.

The so-called bitter and sweet species occur in separate areas of the Americas, suggesting that they were domesticated at different locations. (There is no sharp distinction between bitter and sweet cassavas, though sweet types usually, but not always, have lower levels of prussic acid than the bitter types.) Some scholars suggest that bitter cassava may have been domesticated in the northern part of South America, probably east of the Andes, and that sweet cassava was independently domesticated in Central America.

After its domestication, cassava spread throughout the tropical areas of the Americas. By the time the Spanish conquistadores arrived, cassava was a dietary staple on the islands of the Caribbean as well as on the continental land mass.

At present, more cassava is grown outside its original area of domestication than within. The crop was introduced to the Congo basin more than 400 years ago—records of the rich Portuguese trade between Brazil and the west coast of Africa include evidence that cassava was taken to the Congo basin as early as 1588—and more recently, to Asia and Oceania. It is now grown throughout the tropics. There is no overall pattern of introduction and spread from a few centers of introduction; there have been a multitude of points of entry. Cassava was introduced on several occasions through different ports on the west coast of central Africa, and it spread rapidly throughout what is now Angola, Zaire, Congo, Gabon, and Cameroon.

Cassava had been independently introduced into East Africa and Madagascar by the middle of the eighteenth century. A hundred years later, it was a staple on the coastal strip and around Lake Tanganyika. The cassava grown near Lake Tanganyika comes from introductions to both the west and east coasts. Cassava is now an important element in the diet of most areas of the lowland tropics of East Africa and is a staple in Tanzania, Mozambique, and Madagascar.

Cassava spread throughout West Africa, and with the return of freed slaves from Brazil, the traditional technology for the production of *farinha,* a meal made from cassava, arrived. In Africa, the process was modified to that which is now used for the production of *gari.* Since 1900, the production of both cassava and *gari* has spread rapidly, and cassava is now a staple throughout the wetter regions of West Africa.

The introduction of cassava into Asia is not well documented. The Portuguese may have brought cassava to Goa on the Indian subcontinent in the early eighteenth century. Cassava was probably introduced to Indonesia and the Philippines from Mexico at about the same time. There is evidence that cassava was taken from Indonesia to Mauritius about 1740, which indicates that it was in Southeast Asia before this date. Sri Lanka officially imported cassava cuttings from Mauritius in 1786, and India made importations directly from South America in 1794 and from the West Indies in 1840. The first official record of cassava introduction into Malaysia is in 1886 through the Singapore Botanic Gardens; however cassava was widely grown in Malaysia before that, probably after introduction from India or Indonesia.

By the nineteenth century, cassava was firmly established as a crop in South and Southeast Asia. In Malaya, the Chinese had developed a trade in tapioca (as cassava is called in much of Asia) with Europe, and by 1906, 100,000 hectares had been planted to cassava. Cassava was also an important staple in Indonesia, being used as a rice substitute in Java. Cassava is now grown throughout South and Southeast Asia and southern China: as a staple food in southern India, parts of Burma, and Indonesia; both as an industrial and a subsistence crop in the Philippines, Malaysia, and Taiwan; and as an export crop in Thailand.

Adaptation

The cassava crop is limited to the hotter areas of the world; however within this zone, it is cultivated under enormously differing climatic and soil conditions. For example, in Colombia, a country with diverse ecological conditions, cassava grows in the high-rainfall areas of the Andes at 2000 meters, in the semi-arid areas of the Guajira, on the rich soils of the Cauca and Tolima valleys, on the acid-infertile soils of the savannas in the eastern plains, and in the tropical rain forest of Putamayo. Although different varieties and agronomic practices are used, the fact that the species can be grown under these varied conditions demonstrates its broad adaptability.

Climate. Cassava is grown almost exclusively in the hotter lowland tropics and is never grown as a crop further from the equator than 30°N or 30°S. In the more extreme latitudes where cassava is grown, light frosts occasionally occur. Frosts may

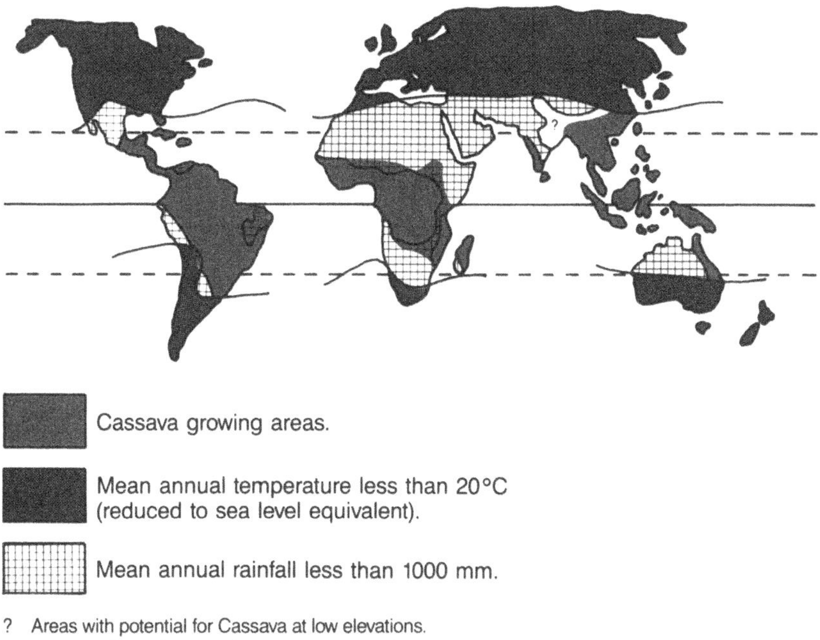

Figure 1. The limits of cassava growing areas are determined largely by rainfall and temperature.

defoliate the plant, but when warmer weather returns, the plant sprouts from the base and grows normally. In areas that have large seasonal variations in temperature, cassava is not grown if the annual mean temperature is less than about 20°C (see Figure 1). In highland areas near the equator (for example, the Andean zones of Colombia, Ecuador, and Peru), cassava grows in areas where the mean annual temperature is as low as 17°C, but fluctuations about this mean are slight. Furthermore, the varieties used in these areas have been specifically selected by the local farmers so that they are well adapted. These varieties do not yield well in areas that have higher temperatures, nor can lowland varieties be successfully introduced into highland areas. These rather special cold-tolerant varieties of cassava are found only in the Andean zone. They are grown as high as 2300 meters above sea level and do not appear to have been introduced to other highland tropical areas, where they probably would be successful.

Most cassava is grown in areas where the average rainfall is over 1000 millimeters a year. Cassava is well adapted to rainfall ranging between 1000 and 3000 millimeters a year, but it needs good drainage. On heavy soils, one day of flooding can destroy the crop. (There are several unconfirmed reports that cassava varieties from the Amazon jungle and the rain forest on the Pacific coast of Colombia are able to survive several weeks of standing water.)

Although cassava does not withstand flooding or even very moist soil conditions over prolonged periods, it is highly tolerant of drought, and cassava is a major crop in areas that have dry seasons as long as 6 months. At the onset of dry weather, the cassava plant reduces its leaf area by producing fewer new leaves while continuing to shed older leaves. The stomata of the remaining leaves close partially, which diminishes the plant's transpiration rate and conserves water. If the dry period continues, more leaves are shed, lowering the leaf area to minimal levels, and both root growth and top growth cease. The plant becomes essentially dormant. When the rains resume, the plant draws on carbohydrate reserves in the stems and roots to produce new leaves, and the plant again becomes productive (see Figure 2).

Once established, cassava, unlike other crops, has no critical period when lack of rain will cause crop failure. A dry spell will reduce yield, but only if it is so extended that the plant dies will crop failure occur. Consequently, cassava is extraordinarily well adapted to areas in which rainfall is uncertain. In addition, because it has no period of acute vulnerability to dry weather, cassava can be planted over a longer time span than most crops, making it easy to fit it into various farming systems.

Cassava can be grown in some areas in which the annual rainfall is less than 1000 millimeters, such as parts of northeastern Brazil and eastern Africa. In these areas, rainfall is also quite variable from year to year, hence it is difficult to place an absolute lower limit on the rainfall in areas where cassava is grown. Nevertheless, cassava is rarely important where the mean annual rainfall is less than 750 millimeters, or where there frequently is less than 600 millimeters of rain a year.

Soils. Cassava is well adapted to the low fertility soils that predominate in the tropics, and it is therefore usually grown on highly weathered and leached soils of the orders Oxisols, Ultisols, and Alfisols. Small amounts are grown on moderately weathered

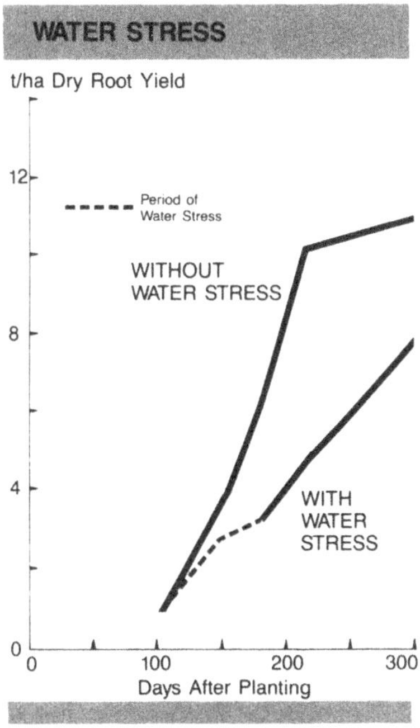

Figure 2. Root yield as affected by a dry period. (Adapted from D. J. Connor, J. H. Cock, and C. Parra. The response of cassava to water shortage: I. Growth and yield, *Field Crops Research* 4:181–200.)

Inceptisols (particularly in India) and on sandy Entisols. Cassava is also grown on a tremendous range of other soils; however the area is insignificant by comparison.

When grown in low fertility soils, the total growth of cassava suffers less than that of most other crops. Similarly, for root yield, which is what farmers are interested in, cassava is an efficient producer in low fertility soils. In an experiment in Colombia involving the application of phosphorus to an infertile soil, scientists found that the higher the rate of phosphorus applied, the higher the total weight of the cassava plant. But root yield was raised only slightly with phosphorus application levels greater than half the level required for maximum plant growth. Thus, when grown in poor soils, cassava may not reach its full potential for total biomass production, but a larger proportion of the total dry matter is found in the roots.

High concentrations of salts or high soil pH have more severe effects on cassava than on most other crops. On the other hand, cassava grows remarkably well where there is low pH and associated high aluminum levels, which are characteristic of many of the well-drained tropical soils on which cassava is widely planted.

Cassava's tolerance of soil acidity is shown by the fact that liming soils that have pH levels as low as 4.4 has no effect on yield, provided that the aluminum levels are not excessive. But even high levels of aluminum barely affect yield. When cassava is grown under experimental conditions in solutions with various concentrations of aluminum, its yield is depressed only by very high levels, while in comparable experiments, the yield of most other crops drops sharply as aluminum levels increase.

In the field, no reduction in cassava yield has been observed in soils that have aluminum levels as high as 80 percent saturation. In the Oxisols of the eastern plains of Colombia, where crops such as maize, sorghum, rice, and beans will not produce any harvest, cassava will sprout, grow, and yield modestly (5 to 6 t/ha) without soil amendments.

The growth cycle of a typical cassava crop is close to a year. The roots start bulking about 3 months after planting (see Figure 3) and continue to increase in weight until 9 to 15 months after planting, when the crop is usually harvested. In other words, the cassava plant accumulates carbohydrates in its useful plant parts—the roots—for 75 percent of its overall growth period.

This pattern is quite different from that of cereal crops. Cereals have a long growth period during which the leaves, stems, and inflorescences develop, followed by a short period during which the carbohydrate accumulates in the grain. A tropical rice crop, for example, reaches maturity in about 4 months. Grain filling occurs only during the final month, so the plant stores carbohydrates in the useful parts, the grain, during only 25 percent of the growth cycle. The contrast in growth pattern is the fundamental reason cassava has a higher yield potential than cereals.

Special attributes

Cassava's special attributes have earned it a reputation as a "famine reserve crop," that is, a crop to be planted in case other

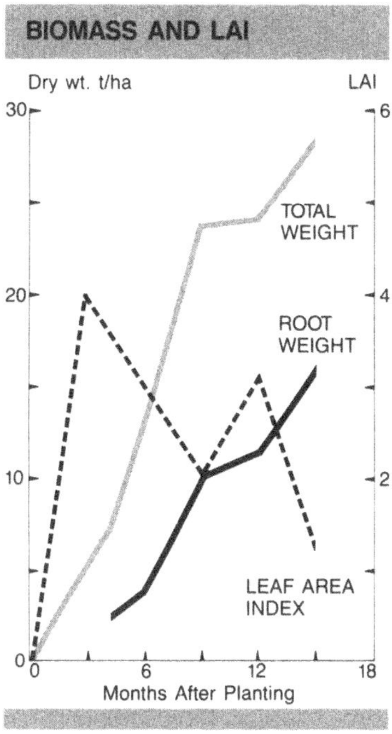

Figure 3. Growth and development of a typical cassava crop. Leaf area index (LAI) is a measure of the leafiness of crops.

crops fail. In colonial times, district officers in Africa insisted that a certain amount of cassava be planted as a famine reserve. More recently, during the Nigerian civil war when widespread starvation occurred, cassava was planted as a basic staple. In transmigration programs in Indonesia and South America, when colonizers arrive, one of the first acts of the newly settled communities is to plant cassava to ensure a food supply in the risky first years.

Cassava can serve as a famine reserve for several reasons, aside from its tolerance of drought and infertile soil. One is that its harvesting date is highly flexible. Most other crops have to be harvested within a specific period—if the harvest is late, most crops spoil or fall to the ground and then germinate or rot. Cassava, on the other hand, can remain in the ground until it is needed. It continues to grow, though the roots may become more fibrous, which lowers the eating quality. Cassava may be left to grow for 2 to 3 years before it is harvested, or it can be partially harvested as food is needed. That is, one or two roots may be harvested but the plant left standing so that it continues to grow.

In Africa, plagues of locust have made famine reserve crops extremely important: Not only is cassava continuously available because it has no fixed harvest date, it is also highly resistant to locust attacks. Even if locusts eat the foliage, the roots remain unharmed, and new shoots rapidly form with little yield loss. In areas in which hurricanes or typhoons frequently occur, cassava is also an important "insurance crop." As long as the plants are neither flooded nor uprooted during such storms, hardly any damage is done. Loss of foliage and the breaking of stems and shoots will not kill the crop.

2
Food, feed, and industrial uses

Cassava is grown chiefly as a food, but it is also an important animal feed, and it has several significant industrial uses. Aside from the economic activity inherent in cassava growing itself, the processing and marketing of cassava generate employment and provide opportunities for the development of rural industries.

As one of the cheapest sources of food energy, cassava should play a major role in meeting developing countries' rising consumption of both food and animal feed. The inability of most countries to expand cereal production fast enough to meet the demand often forces them to spend scarce foreign exchange to import cereals. Cassava can help fill the gap by enlarging total food supplies because production is possible on land that is not well suited to cereals and other food crops. Moreover, growing cassava to provide carbohydrate for animal feed means that better land that is devoted to the production of feed grains can be switched to food production. Developing countries already consume 12 million tons of cassava annually for animal feed, but the potential is much greater. Industrial uses, mainly commercial starch, consume about 5 percent of the total cassava production. Recent developments suggest that cassava has promise as a source of renewable energy.

FOOD

Cassava is often regarded as a poor food with low nutritional value that is only consumed by subsistence farmers who have little else to eat. Actually, cassava is an excellent source of dietary

Table 5
Dry matter, carbohydrate, and protein content of root and tuber crops

Crop	Dry matter (%)	Carbohydrate (% dry matter)	Protein (% dry matter)
Cassava	37.5	92.5	3.2
Irish potato	22.0	85.9	9.1
Sweet potato	30.0	91.0	4.3
Yams	27.6	87.3	8.7
Taro	27.5	84.4	6.9

Source: K. A. Leslie, The significance of root crops in the tropics, in *Proceedings of the International Symposium on Tropical Root Crops,* 5:1–12 (St. Augustine, Trinidad: University of West Indies, 1967).

energy, and it is shortsighted to consider cassava solely a subsistence or famine crop.

Nutritional value

Cassava roots have 30 to 40 percent dry matter, which is a higher proportion than most other roots and tubers have (see Table 5). The dry matter content depends on such factors as the variety, the age of the root at harvest, the soil and climatic conditions, and the health of the plant. Starch and sugar are the predominant components (approximately 90 percent) of the dry matter; starch is by far the most important. The metabolizable energy of dry cassava, 3500 to 4000 kcal/g, is similar to that of maize flour.

The crude protein content of cassava roots, normally estimated by multiplying the nitrogen content by the factor 6.25, is 2 to 3 percent on a dry matter basis. But the true protein content is less, because up to half the nitrogen in the root is nonprotein nitrogen. The quality of the protein is reasonably good, though sulfur amino acids are deficient. Concern with protein quality is, however, somewhat academic because the protein level is so low in fresh roots, and is further decreased by processing to *gari* or *farinha,* that it has little nutritional value.

Table 6
Food composition of 1 kilogram of cassava compared with daily requirements for an adult male

	Cassava[a]	Adult male daily requirement
Food energy (cal)	1460	2500
Water (g)	625	
Carbohydrate (g)	347	
Protein (g)	12	65[b]
Fat (g)	3	
Calcium (mg)	330	500
Iron (mg)	7	8
Vitamin A (IU)	trace	2500
Thiamine (mg)	0.6	1.2
Riboflavin (mg)	0.3	1.2
Niacin (mg)	6	15
Vitamin C (mg)	360	25

Source: K. A. Leslie, The significance of root crops in the tropics, in *Proceedings of the International Symposium on Tropical Root Crops,* 5:1–12 (St. Augustine, Trinidad: University of West Indies, 1967).

[a] Fresh cassava of relatively high dry matter content.

[b] Current estimates of protein requirements are considerably less than this figure. WHO/FAO suggests 46 g of protein of 80 percent quality relative to milk or eggs.

The roots contain significant quantities of vitamin C, thiamine, riboflavin, and niacin (see Table 6). A person obtaining more than 250 cal/day from cassava would satisfy the daily vitamin C requirement. But boiling reduces the vitamin C content by 50 to 70 percent, and processing to such products as *farinha, gari,* and *fufu* reduces it by 75 percent or more. Nevertheless, in areas in which cassava consumption is high, even processed cassava can supply sufficient vitamin C to satisfy minimum daily requirements.

In nutritional terms, cassava must be considered primarily as an energy source that contributes little else, except vitamin C, to the nutrition of the people who eat it. Cassava has been criticized and expansion of its production sometimes opposed because of its low nutritional value. Cassava has also been implicated as a major cause of malnutrition in parts of Africa.

It is of course true that people who consume large quantities of cassava with little protein supplement will be malnourished. Cassava does, however, have the tremendous advantage of being cheap, so it can be used to satisfy the energy requirement of a diet at low cost, releasing money for the purchase of more expensive high-protein foods. It has been conclusively shown that increasing the percentage of protein in diets that are deficient in total energy will not alleviate protein deficiency because an undernourished body partially utilizes protein in the diet as an expensive energy source. On the other hand, it has been demonstrated that increasing the carbohydrate supply in the diet can increase the efficiency with which protein is utilized. Thus the nutritional role of cassava is as a cheap energy source, with other foods providing the necessary protein, vitamins, minerals, and fats.

Toxicity

Raw cassava contains the glycosides linamarin and lotaustralin, which are converted to hydrocyanic, or prussic, acid, a poison, when they come in contact with linamarase, an enzyme that is released when the cells of cassava roots are ruptured. Although occasionally deaths from consuming raw cassava roots have been reported, the traditional processing and cooking methods reduce the cyanide levels. If normal preparation procedures are used, acute cyanide toxicity does not occur. Chronic cyanide toxicity occurs in some localities in Africa where cassava consumption is high, up to the equivalent of 1 kilogram or more of fresh roots per day over a long period, and where the consumption of iodine and protein, particularly animal protein, is extremely low. In Nigeria and Zaire, ataxic neuropathy (nervous degeneration) and goiter (which leads to cretinism in severe cases) have been associated with high levels of cassava consumption.

When cyanide enters the bloodstream, it is converted to thiocyanate, a sulfur-containing compound, by the enzyme rhodanase. The thiocyanate is later excreted in the urine. The compound plays its toxic role by using up body sulfur in detoxification, thus increasing the body's demand for sulfur-containing amino acids, or by interfering with the iodine uptake of the thyroid, resulting in goiter. In both cases, high cassava consumption aggravates problems associated with low levels of sulfur amino acids and iodine in the diet.

Chronic cyanide toxicity has not been reported in areas of high cassava consumption in Latin America or Asia, which reinforces the hypothesis that goiter and ataxic neuropathy are caused by a complex interaction of several factors. Of course in Latin America, cassava consumption levels are generally much lower than in Africa, but in studies in the Amazon jungle, where tribal people eat as much as 1 kilogram of cooked fresh cassava per day per person and wash it down with 3 liters of fermented cassava beer, *masato,* there have been no reports of either goiter or ataxic neuropathy. (The symptoms of goiter are so obvious that it is unlikely to be missed.) These tribes consume considerable quantities of animal and fish protein and thus have high levels of sulfur amino acid intake.

Kerala, India, is another area that has high cassava consumption without adverse effects. Cassava provides more than 800 calories per adult equivalent per day, and the total protein intake is low—41.5 grams per adult equivalent per day—yet there are no reports of goiter or ataxic neuropathy related to high cassava consumption in this area. Moreover, among the states of India, Kerala has one of the lowest rates of infant mortality and the longest life expectancy, indicating that high levels of consumption of properly prepared cassava need not be detrimental to health.

There is little doubt that high cassava intake may aggravate the effects of low protein levels in the diet, particularly when sulfur amino acids are deficient and iodine intake is low. The solution to the problem seems to lie with the use of iodine, probably as iodized oil, and the consumption of sufficient animal or fish protein. The requirements for iodine and sulfur amino acids are slightly increased when cassava consumption is high.

Beneficial effects

Much has been written about the deficiencies of cassava as a food, but little has been reported on its possible beneficial effects on health. For example:

- In the treatment of sickle-cell anemia, which is most common in people of African descent, thiocyanate has been used. Since the blood levels of thiocyanate are increased by ingesting cassava, it may be an important means of alleviating the disease in Africa.
- Schistosomiasis is an extremely widespread and debilitating disease common in the lowland tropics. Amygdalin, a cyanogenic glycoside similar to linamarin, has been suggested as a medication for schistosomiasis, and it has been proposed that cassava in the diet has similar effects.
- The possible role of cyanogenic glycosides to control cancer is extremely controversial; nevertheless it has been suggested that some cancer cells are unable to detoxify cyanide as they do not possess the enzyme rhodanase. Clinical tests with amygdalin failed to show any effectiveness in controlling cancer.

These possible beneficial effects of cassava are based on few results and much speculation, and they need to be more thoroughly researched before any conclusions can be reached about their real importance.

Cooking and processing

Cooking and processing offer some solution to two of the drawbacks of cassava—its toxicity and its perishability. Although raw sweet cassava is occasionally eaten as a snack or thirst quencher in the Congo region, Tanzania, and West Africa, cassava is not generally consumed raw because of its toxicity. All forms of cassava processing lower the levels of cyanogenic glycosides and prussic acid in the final product. The degree of reduction varies greatly with the type of processing. Processing cassava into a dried product also is a way of dealing with the perishability of cassava because, once harvested, fresh cassava roots show deterioration within as little as 24 hours.

Cooked fresh cassava

The simplest way of preparing cassava is to peel it immediately after harvest and boil it for 10 to 40 minutes until it is soft.

The boiling time needed depends on the variety. After boiling, the cassava may be eaten, or it may be cut into long chunks and deep fried. However, sweet cassava must be used. If bitter cassava is used, the unpleasant taste remains after cooking, and the food is dangerous as well. Although boiling destroys the enzyme linamarase and drives off the prussic acid, the linamarin itself is not destroyed, and a long-term ingestion of it may lead to chronic cyanide toxicity in people whose diets lack sufficient protein and iodine.

Fresh cassava, which is often cooked in a soup or gruel, is eaten throughout the cassava growing areas of the tropics. Consumption is much greater in rural areas than in cities because of the perishability of cassava. The various flours and meals keep better than fresh cassava.

Cassava meal

The forms of cassava meal can be roughly divided into unfermented and fermented meals. The former are made by grinding or slicing peeled roots and then drying and milling them to form a meal. *Gaplek* in Indonesia is made from dried chunks that are ground; *kokonte* in Ghana is prepared in the same way except the chunks are not ground immediately after drying. In Brazil, large processing plants mechanically wash, peel, and grate the cassava, which is then artificially dried using wood-fired or oil-fired furnaces before milling to produce *farinha de raspa*. In Africa, the cassava is often soaked in pools of water for several days until the roots soften; they are then peeled and sun dried before being ground to produce a meal.

Farinha d'agua and farinha grossa

The production of *farinha* is a traditional process to produce a meal from bitter cassava. The processes for making *farinha d'agua* and *farinha grossa* are essentially the same except that in the former, the roots are submerged in pools of water for several days before peeling. The peeled roots are grated by hand in the traditional process, but mechanically nowadays. The grated roots are then stuffed into a tipiti, a long, thin basket-weave tube. When the tipiti is stretched, its internal volume is reduced,

A tipiti, a basket-weave tube, being used to squeeze the water out of grated cassava roots, and in the process, eliminating much of the prussic acid. (*Source:* CIAT)

squeezing liquid from the mashed roots. The liquid contains some starch and may be used as a base for stews and soups, or the starch can be recovered by allowing it to settle and removing the liquid. The moist mash from the tipiti is placed in large, flat pans, usually made of copper, and roasted until dry. The dry meal can be packed in bags and stored.

Gari

The production of *gari* is a modification of the process for the manufacture of *farinha* in which fermentation and cooking affect the flavor. Roots are washed, peeled, and grated. The resulting mash is placed in bags and squeezed by heaping stones and logs on the bags, a process that is not nearly as effective as the tipiti. The bags are left for several days, during which time the mash ferments. The fermented mash is then roasted or fried, often with palm oil, in small pots until dry. Then it may be packed in bags and stored. In the markets of West Africa, an enormous array of different kinds of *gari* can be found.

Chickwangue

Chickwangue is an African product that has no direct equivalent in Latin America. The cassava is soaked in water for several days until it softens, and the root is then peeled and macerated. Fibers are removed, and the resulting paste or dough is wrapped in banana leaves. It may be eaten with no further processing, or after boiling.

Other food products

Many other products can be made from cassava, including beer, biscuits, and cakes. *Casabe* and *fufu* are among the more important products. In the Caribbean islands and on the Atlantic coast of Colombia and Venezuela, cassava is processed in the same manner as for the production of *farinha,* except that the squeezed cassava mash is kneaded into a flat cake that is baked to form a tortilla-like bread called *casabe.* The name "cassava" originates from this product. In Africa, *fufu* is made by taking the mash used for making *gari* and fermenting it underwater and then drying it.

Employment in cassava processing

Most of the world's cassava is produced as a dietary staple of the poor, and approximately half of the cassava used for food is eaten after specialized processing, usually by the family or on the village level. The labor used in processing cassava probably is greater than that used in growing and harvesting it. Traditional processing methods are so laborious that the number of mechanized village processing units is increasing in many regions. The equipment used in village plants is very simple and, except for the motors, locally made, thus creating a stimulus to small

In the Caribbean islands, cassava mash is baked to form *casabe*, which can be stored for long periods. (*Source:* CIAT)

industry. Even the processing of cassava chips or pellets and cassava starch for export is done to a considerable extent in small units, although large ones are common for pelleting and starch extraction.

The simplest processing technique is to peel the roots and dry them on racks or trays in the sun. This work is commonly done by family members in Zaire, India, Indonesia, and Tanzania. In Zaire, the processing of cassava and storage require as much as 20 man-days of labor per ton of fresh cassava.

The labor needed to produce specialized meals such as *farinha* and *gari* can be enormous. Most of these meals are made on the farm or in small village units. In East Central State of Nigeria in the mid-1970s, the majority of the farm families processed the cassava by hand; only 24 percent sent the cassava to be grated mechanically in the village units. Hand grating of cassava, alone, requires some 10 to 15 man-days per ton of fresh cassava. In addition, the cassava must first be peeled and then pressed to remove the water, roasted, and packed. The use of small mechanical graters greatly reduces the labor needed for processing. In villages in Brazil, small mechanized units are common.

Large processing factories are being established in areas that have a plentiful supply of cassava. Much of Brazil's cassava flour, which is used as a partial substitute for wheat in bread, is produced in factories that mechanically wash and grate the cassava and then dry it in furnaces before milling and packing. Large mechanized *gari* plants are being constructed in Nigeria and Ghana. Since they need 35 to 40 tons of fresh cassava per day, they normally are linked to large plantations.

Marketing cassava

The marketing of dried cassava products such as *farinha, gari,* and cassava flour can easily utilize the normal channels for staples such as rice or maize. Fresh cassava, however, must be marketed rapidly before it deteriorates. With such a perishable product, arbitrage is not possible over long distances, and hence sharp, sudden price fluctuations might be expected. That they do not generally occur suggests that highly sophisticated marketing systems have been developed.

Fresh cassava must be sold to the consumer within 1 to 2 days after harvest, and it must be eaten in less than a week. In rural areas or areas close to urban centers, the cassava is harvested and taken by a family member directly to the local market for sale. In Kerala, India, for example, it is common to see women carrying baskets of cassava on their head to the local marketplace, where they may sell it directly to the consumer or to a shopkeeper or stall owner. This marketing system is common throughout the tropics where the production site is near the market. Because of close contact between the producer and the consumer, the farmer can regulate harvesting to keep the supply in balance with the prevailing demand.

In areas in which the markets are far from the production areas, more sophisticated marketing systems have evolved. In

West Africa and parts of South America, fresh cassava may be transported long distances by intermediaries to the populous urban centers. These intermediaries provide transport and regulate the flow of cassava to the markets. The intermediary may either contract farmers to harvest a certain amount of cassava for a particular day, which he then collects, or he may buy a field of cassava to harvest with his own labor according to market demand. The difference in price received by the producer and that paid by the consumer can be considerable, but the person who deals with such a highly perishable commodity has to accept a high level of risk. He cannot store the product if his transport fails or if the market is saturated, so he may be left with a worthless load of rotting cassava.

Future demand for cassava as food

The world's population will double in 50 years, and most of this increase will occur in developing countries, which are concentrated in the tropics. The International Food Policy Research Institute has estimated that to meet market demand, global food production must increase at approximately 3 to 4 percent a year and, if the minimal nutritional requirements of the world population are to be met, food production will have to increase by 5.4 to 6.3 percent a year. The inability of developing countries to keep pace with the increasing demand for food and feed forced them to double their cereal imports to 100 million tons during the 1970s.

National planners and the people who administer international agricultural programs have focused on cereals research to increase food production. Based on the current importance of cereals in national food budgets, this is a sound strategy. However, this policy has tended to promote the idea that cereals are *the* solution, rather than one potential solution among many. Thus, for example, 1975 research expenditures as a percentage of total value of production in Asia were much greater for cereals than for roots and tubers (see Table 7). Within the system of international agricultural research centers, the support for cassava in 1982 was only 2 percent of the total budget, while 5 percent of the

Table 7
Developing countries' expenditures for research on starchy staples, 1975

Commodity	Value (million US$)	Research expenditure (million US$)	Research expenditure compared with value of product (%)
Sorghum	1500	12	0.77
Maize	3000–4000	29	0.75
White potatoes	1000	8	0.68
Wheat	5000–6000	35	0.65
Sugarcane	5000–6000	30	0.50
Rice	Over 13,000	34	0.26[a]
Sweet potatoes	3000–4000	3	0.09
Cassava	5000–6000	4	0.07

Source: Adapted from supporting paper to the "World Food and Nutrition Study" (Washington, D.C.: National Academy of Sciences, 1977), 5:52.

[a] Shallow water rice, 0.40.

total calories in the food-deficient countries came from cassava. The bias toward cereals has tended to obscure the possible role of root crops in alleviating the food and basic energy shortfalls of the less-developed countries.

One reason for the lack of attention to root crops, particularly cassava, is the generally held belief that they are not preferred food, so that as living standards rise, demand and consumption will fall. Hence it is argued that the importance of cassava as a food crop will diminish as countries develop. Although this scenario may have some validity for cassava meal, it is by no means true of fresh cassava and many other cassava products.

Fresh cassava

Studies in Indonesia have shown that per capita purchases of fresh cassava roots tend to increase as the income level increases and then remain stable at higher income levels (see Table 8). Similar data have been obtained in Brazil, where the income elasticity of demand for fresh cassava was found to be positive, and in Ghana, where there was no tendency for purchases of

Table 8
Consumption of fresh cassava and cassava flour (*gaplek*) by income class in Java, 1976

Income class (rupiahs/person)	Consumption (g/person/week)	
	Fresh roots	*Gaplek*
Less than 1000	203	142
1000–1999	318	283
2000–2999	414	215
3000–3999	527	174
4000–4999	664	123
5000–5999	662	78
6000–7999	640	60
8000–9999	640	30
10,000 and above	607	35

Source: W. O. Jones, Cassava in Indonesia: Preliminary observations, mimeograph, Food Research Institute, Stanford University (Stanford, Calif., 1978).

fresh cassava to decline as the income level increased (see Table 9). However, similar trends may not hold in urban Nigeria. Furthermore, in Indonesia, the cross-elasticity of demand of cassava with rice was very high, suggesting that if the introduction of better production or storage techniques reduced the price to the consumer, cassava consumption would increase substantially.

Fresh cassava is mainly marketed in rural production areas or in urban centers near production areas, but this fact does not necessarily indicate a low demand in other areas. In Colombia, cassava cannot be grown near Bogota, the largest urban center, because the climate of the surrounding farming areas is too cold. Nevertheless, there is a substantial market for cassava in Bogota, and prices may be up to threefold those received by the farmers. In addition, there is a price premium of 40 to 60 percent for high quality cassava. These observations indicate that fresh cassava of good quality is a preferred food, at least in Colombia.

Fresh cassava is often the cheapest source of calories in rural areas (see Table 10). However, because of the high marketing margin, fresh cassava is often a luxury item in urban centers. The bulkiness of the crop and its extreme perishability make it expensive to transport and risky to handle. Cassava roots start

Table 9
Consumption of fresh and dried cassava (*kokonte*) by income class in Ghana

Monthly household income (Ghana shillings)	Cassava consumption (cal/person/day)	
	Fresh	*Kokonte*
100–199	482	138
200–249	402	56
250–299	491	46
300 and above	453	10

Source: H. Kaneda and B. F. Johnston, Urban food expenditure patterns in tropical Africa, *Food Research Institute Studies* 2 (1961):229–275.

to deteriorate within 24 hours after harvest, thus the middleman has to buy and sell the cassava quickly. In addition, by the time fresh cassava arrives at the urban markets, it has often already started to deteriorate, so consumers are forced not only to buy a low quality product but also to eat it immediately after purchase (except for the affluent who can refrigerate it). Furthermore, because low starch content is normally associated with longer shelf life, producers for the urban market are likely to select varieties that have a low starch content. But low starch content is related to low quality. Thus, because of the high perishability of cassava, the urban price is high, the quality is low, and consumers who do not have refrigerators cannot afford to eat cassava on a daily basis. Taking steps to improve the shelf life of cassava would reduce the consumer price (assuming a downward adjustment in marketing margin), improve quality, and expand demand in urban areas.

Cassava flour

Cassava flour is an extremely cheap source of calories (see Table 10), but it is not a preferred food. Studies on *kokonte* in Ghana and *gaplek* in Indonesia show that consumption decreases as the income level increases (see Tables 8 and 9). Researchers in Brazil and Nigeria have found that as income rises, there is

Table 10
Prices of cassava relative to prices of other staples

Country	Year	Crop compared	Cassava form	Calories of cassava that can be purchased for same amount as 1 calorie of compared food
India	1970/71	rice	not stated	2.23
Indonesia	1976	rice	*gaplek*	2.27
		rice	fresh	1.51
		maize	*gaplek*	1.32
		maize	fresh	0.87
Ghana	1955	rice	*kokonte*[a]	3.18–1.96
		rice	fresh	1.94
		maize	*kokonte*[a]	1.12–1.33

		maize	fresh	0.81–1.11
Nigeria	1973	maize	*gari*	1.08
		rice	*gari*	2.42
Nigeria	1975/76	maize	fresh	1.23
Brazil	1975	rice	*farinha*	2.10
		rice	fresh	0.91
		maize	*farinha*	1.75
		maize	fresh	0.76

Sources: India: United Nations, *Poverty, unemployment, and development policy: A case study of selected issues with reference to Kerala* (New York, 1975); Indonesia: W. O. Jones, Cassava in Indonesia: Preliminary observations, mimeograph, Food Research Institute, Stanford University (Stanford, Calif., 1978); Ghana: T. T. Poleman, The food economies of urban middle Africa: The case of Ghana, *Food Research Institute Studies* 2 (1961):121–175; Nigeria 1973: J. Goering, *Tropical root crops and rural development* (Washington, D.C.: World Bank, 1979); Nigeria 1975/76: W.N.O. Ezeilo, Intercropping with cassava in Africa, in *Intercropping with cassava: Proceedings of an international workshop held at Trivandrum, India, 27 Nov.–1 Dec. 1978,* ed. E. Weber, B. Nestel, and M. Campbell, 49–56 (Ottawa, Canada: IDRC, 1979); Brazil: Fundacao Instituto Brasileiro de Geografiae Estatistica, *Estudo nacional da despesa familiar* (Rio de Janeiro, 1978).

[a] Dried roots.

a tendency for the consumption of dried cassava to fall and that of fresh cassava to increase. In general, throughout the developing world, the consumption of cassava flour declines as income rises. Urbanization also has an effect. With the possible exception of parts of West Africa, the consumption of cassava flour tends to be lower in urban areas than in rural areas. Since the proportion of the population living in cities is rising in most nations, the demand for traditional cassava flour products is likely to fall. Consequently, population growth may raise the total demand for cassava flour, but not as fast as the rate of population increase if income levels rise and a greater proportion of the total population lives in the cities. The net effect of these crosscurrents will be relatively little change in the overall demand for traditional cassava flour products in areas in which the population increase is large and development slow, but in countries with low population growth and rapid development, demand may fall sharply.

The situation may be different if a large portion of the population is extremely poor. As incomes rise among the extremely poor, consumption of cassava flour tends to expand, which suggests that if incomes increase, or cassava prices decrease, consumption could grow substantially among the very poor. Furthermore, cassava demand depends on the relative price of other food staples. In Indonesia, where the staple food is rice, increases in the price of rice lead to a higher consumption of *gaplek*.

These observations suggest that if new technology were developed to substantially lower the price of cassava flour relative to other staples, consumption would increase. This result is only likely to occur when people are so poor that they buy the cheapest calories, regardless of the form in which they are available. It is nevertheless important not to underestimate the potential role of cassava as the cheapest source of calories to alleviate the caloric deficits of the large number of abjectly poor in the lowland tropics.

Novel uses of cassava flour

In most developing countries, the demand for bread is increasing rapidly. Since few of these countries can successfully grow wheat on a large scale, most of them import wheat and sell it at subsidized prices. This policy not only strains foreign reserves and is costly to the national treasury, it also discourages the use of locally produced substitutes.

Many studies have shown that cassava flour or starch can readily be incorporated into bread at levels as high as 10 to 20 percent with little decline in quality. The addition of stearyl lactylate or similar additives is necessary to strengthen the dough. Although mixing cassava flour with wheat flour will lower the protein content of the resulting loaf, that disadvantage can easily be overcome by fortifying the cassava flour with soybean flour.

Much of the research on the incorporation of cassava products into bread has been done with cassava starch, and levels of up to 40 percent have been incorporated while maintaining acceptable loaf quality through the use of extenders. Cassava starch is more expensive than cassava flour because of the high costs of processing, so it is preferable to use cassava flour. Bread made from cassava flour is of lower quality than bread made from cassava starch mixtures, but nevertheless, levels of cassava flour as high as 30 percent have been successfully incorporated, and 10 to 15 percent incorporation produces a loaf that is virtually indistinguishable from a pure wheat loaf. In northeastern Brazil and Senegal, up to 25 percent cassava flour is frequently used in bread. The feasibility of incorporating fresh cassava into wheat flour for bread making, thus eliminating the drying process, has been demonstrated by the Tropical Products Institute (TPI) in the United Kingdom. TPI has produced good quality bread by replacing 20 percent of the wheat flour with fresh cassava.

If government policymakers fostered the incorporation of cassava into bread mixes and eliminated subsidies for imported wheat, a large increase in the demand for cassava flour (and possibly fresh cassava) seems likely to result. In 1979, Brazil announced a plan to gradually remove all special subsidies on wheat imports to encourage the use of local products and to reduce the drain on foreign exchange.

Aside from bread, cassava flour can be used in many other products, such as cakes and cookies. In India, cassava goes into spaghetti and noodles. In Indonesia, *gaplek* flour is sometimes used to extend wheat flour for noodles, and starch is also used as a food product, particularly to produce a snack called *krupuk*. In Thailand, a major domestic use of cassava is for confectionery.

FEED

Cassava has become an important animal feed in the last 20 years. The most visible evidence is the export of cassava chips from Thailand, mostly to the European Community. Thai exports rose from 150,000 tons in 1958 to 6 million tons two decades later. Of the estimated 25 million tons of fresh cassava that went into animal feed in 1978, about half was consumed in developing countries, mainly in the Americas. Brazil, whose total production for animal feed is estimated officially at 9 million tons but is probably closer to 4 million tons, is by far the most important tropical user of cassava in animal feed.

Cassava-based rations

Early research on cassava roots as an energy source for livestock showed that pigs and poultry that were fed cassava, even with supplemental proteins, had lower growth rates and feed conversion efficiencies than animals on cereal-based diets. More recently, methionine supplementation has been found to be important. Methionine is an essential amino acid, and it also plays a role in the detoxification of cyanide. With the addition of protein, methionine, and adequate levels of minerals and vitamins, low levels of cassava can replace maize in pig and poultry diets with no decrease in performance. At higher levels, problems may be encountered in getting animals to consume adequate quantities. The powdery nature of cassava and its tendency to form a paste in the animal's mouth makes it difficult to eat. These problems can be largely overcome by pelleting cassava.

In carefully balanced, pelleted rations, up to 60 percent cassava meal can be fed to broiler chickens and up to 50 percent to layers. For fattening pigs from 17 to 35 kilograms liveweight, up to 50 percent cassava can be used in the diet; for heavier pigs, the level can be increased to 70 percent. These data demonstrate the scope for increasing the use of cassava in animal diets in tropical countries and reducing imports of cereals.

Cassava roots are an excellent energy source for ruminants, whether dried or, in the case of sweet cassava, fresh. Trials in which dairy cattle diets are supplemented with cassava have repeatedly shown increases in milk yields, and cassava can readily be used to replace cereals in the fattening of cattle.

In some areas of Brazil, the whole plant is used as cattle fodder after being ground and dried. With their high protein content (20 to 30 percent), the leaves have been successfully used as a protein supplement in beef cattle diets. Research in Venezuela and Colombia has shown the potential of green stems and leaves from cassava as a high-protein forage crop. Annual yields have been as high as 4 tons of crude protein per hectare.

The use of the whole cassava plant as a complete forage has some drawbacks. In Thailand, the export of dried cassava leaves was considered, but because the leaves are a major source of soil nutrients, their loss would have forced growers to purchase costly fertilizers to maintain soil fertility. Thus the feeding of cassava leaves to animals is likely to be practical only on farms where the manure can be returned to the fields.

Chips and pellets

In Thailand, Malaysia, and Sumatra, Indonesia, the production of cassava chips (small irregular chunks in Thailand, shreds in Malaysia, and a variety of shapes in Indonesia) has increased tremendously over the last 10 years. The Thai industry is by far the largest, and small drying units have been set up in villages throughout the cassava production areas of Thailand. A small unit processes about 16 tons of cassava a day. The roots are loaded by hand, or by conveyers, into small, locally built chippers. The chips are then spread on concrete drying floors by hand, mainly by women, and turned every few hours during the day for 2 to 4 days until they are dry enough to be packed into bags. If it rains before the cassava is dry, the chips are pushed into small piles and covered until the rain stops.

These activities employ many people. One plant in Thailand has 70 workers and processes about 40 tons of fresh cassava per

With cassava chippers and concrete drying floors, dried cassava can be produced at low cost. (*Source:* CIAT)

day. Thus the labor for processing is about 2 man-days per ton of fresh cassava. These figures suggest that Thailand, which annually processes close to 12 million tons of fresh cassava, employs 100,000 people or more in the cassava drying industry. Malaysia has a rather similar system, but there tractors are often used to spread and collect the cassava. One study in Malaysia showed that a unit processing 12 to 15 tons a day of fresh roots employed 16 to 20 people; however, another unit with two tractors processed 18 to 26 tons a day with only 12 people. The use of tractors for spreading chips in Thailand is becoming more common.

In Thailand, the cassava is pelleted after drying. Much of the equipment in the pelleting plants was imported until recently, but most plants now use locally made equipment to produce what are called "native pellets." In this manner, the expanding Thai cassava industry provides considerable employment at the farm level in production and at the village level in the drying and pelleting plants, and it stimulates the development of local machinery fabricators.

Future demand for cassava as animal feed

The fact that the European Community (EC) is the largest consumer of cassava for animal feed is related to the heavy tariffs the EC imposes on imported grains and compound feeds to protect its cereal growers. Under the General Agreement on Tariffs and Trade (GATT), cassava, which was not an important feedstuff at the time the agreement was made, was given a special tariff of only 6 percent, or 18 percent of the levy on barley, whichever is lower. These low tariffs have made cassava chips an economically attractive alternative to cereals for feed compounders in Europe.

In 1980, the cassava export trade was dominated by Thailand, which exported 80 percent of the total, and Indonesia, which exported 12 percent. Thailand has developed a deepwater port with storage facilities specifically for cassava exports, and it realizes considerable economies of scale by shipping in batches of up to 100,000 tons. It would be very difficult for any country to displace Thailand from its dominant position in the export market unless the price of the raw material were substantially lower or the quality substantially higher.

The Thai chipping, drying, and pelleting industry is based on low-cost technology. This situation results in a cheap but poor quality product, which is sometimes adulterated with sand and other substances such as wood chippings. The country's prospects for expanding exports to Europe are cloudy. Because the imports of cassava have decreased the demand for barley, there is considerable pressure on the part of European farmers to raise the tariffs on cassava. Without the special tariffs, cassava might not be competitive. So far there has been no change, but Thailand has agreed to restrict its exports voluntarily, and foreign aid is being made available to develop alternative cropping systems in the main producing areas of northeastern Thailand.

Other countries might be able to enter the cassava trade if better drying methods could enable them to produce a higher quality product to be sold at a premium price and if new markets such as Japan and South Korea were developed. Japan annually imports some 9 million tons of maize and 5 million tons of sorghum, which might be replaced, at least in part, with cassava products. Because Japan has a positive trade balance with many of the world's largest grain producers, it finds it expedient to import grain from them, but South Korea arranged to import 30,000 tons of cassava chips in 1982.

The Soviet bloc also imports vast amounts of cereal grains. In recent years, the USSR alone imported more than 13 million tons of maize a year. The USSR aims to increase meat production, and to achieve this goal, it requires large quantities of low-cost energy for animal rations. This demand has been met by imports from the traditional grain-exporting countries, but at least some of this demand could be satisfied in the future by large imports of cassava chips and pellets. However, as most developing countries are importers of basic energy sources, such as cereal grains, it is probably more sensible for them to satisfy their internal demand before moving into the export trade, which is highly competitive.

Since the 1960s, the burgeoning production of pigs and poultry in developing countries has caused dramatic increases in imports of sorghum, maize, and soybeans to feed to the animals. As the demand for meat continues to rise, developing countries will have to sharply escalate their feed grain imports, expand their production of cereals, or find alternative energy sources for animal rations. Increased imports will further strain the balance of payments in the developing world and will not help nations create wealth or employment within their own borders. Expansion of feed grain production, particularly sorghum, has reduced many countries' dependence on imported cereals for animal feed, but feed grain production frequently uses land and other resources, such as irrigation, that could otherwise be utilized for the production of food. A substitute crop that could be grown with few purchased inputs on land that is unused or of only marginal utility for the production of food crops would be more efficient. Cassava has the potential to fulfill this role. Studies in Colombia have shown that even at yield levels as low as 15 t/ha, cassava can compete with cereals such as sorghum.

Mexico, which imports 4 to 5 million tons of cereal grains per year, of which 1 to 2 million tons are destined for use in balanced diets for swine production, has embarked on an ambitious project to produce cassava on the poorer soils of the southeastern part of the country. This project promises not only to reduce dependence on imported grains but also to be a factor in raising incomes in one of the least-developed areas of Mexico.

Malaysia imports considerable quantities of cereal grains—more than 400,000 tons of maize a year from 1979 to 1981. Increased cassava production is replacing imported cereal grains in that country: In 1970, it was estimated that up to 28,000 tons of chips and pellets were being incorporated into animal rations by the feed millers, and by 1980, over 30 local feed mills were using cassava products in their rations.

INDUSTRIAL USES

Cassava is a common raw material in starch production. Cassava starch has uses in the food industry, for papermaking, as a lubricant in oil wells and in the textile industry, and as the substrate for the production of dextrins, which are used in glues. Between the two world wars, alcohol was produced from cassava in Brazil and Australia. This use declined with the availability of cheap supplies of petroleum products, but there has been renewed interest in producing alcohol from cassava as the price of crude oil has risen.

Starch

The cassava growing areas of the Americas and Asia are dotted with small starch factories, which depend on nearby farms for supplies. The factories are highly labor intensive; however they manufacture only a small fraction of the total starch supplies. These units typically have an output of 1 ton of dry starch per day from 5 tons of fresh cassava and employ five to eight people. The cassava is peeled manually, but the rest of the processing is semimechanized, thus allowing one person to process approximately 1 ton of fresh cassava per day. Due to the low

Small starch extraction plants are commonplace in cassava growing areas of Latin America. (*Source:* CIAT)

quality of the starch, these small factories are finding it increasingly difficult to compete with the large mechanized plants, which are less labor intensive.

From the *ralladeros* of Colombia that produce up to a ton of starch per day to factories in Thailand and Malaysia that produce more than 100 tons of starch per day, the process of starch extraction is essentially the same. The roots are washed, peeled, and finely grated. The liquor containing the starch is separated from the fiber by some form of filtration, and the starch is then separated from the water by sedimentation or centrifugation. After being dried naturally or passed through artificial driers, the starch is milled to a fine powder.

In parts of Colombia and Brazil, a special sour starch is made. Wet starch is allowed to ferment for as long as 20 days, and then it is dried in the normal manner. The starch has a distinctive flavor and is mixed with cheese, eggs, and water to make a traditional bread, *pan de yuca* (literally, cassava bread).

Cassava starch is produced in most of the cassava growing countries of Asia and the Americas and in Togo in Africa. But many of the producer countries also import starch periodically. In 1974, Brazil imported US$1.4 million of starch products from the United States. In the first 9 months of 1976, Indonesia imported 59,000 tons of cassava starch, although it exported starch in the 1979–1980 period. In 1982, Indonesia again imported starch from Thailand. Small quantities of cassava starch are also imported by the Philippines and Guyana. These countries have the possibility of becoming self-sufficient in starch production.

The largest importers of cassava starch are the United States and Japan, though imports by the United States have declined in recent years. The cassava starch market has been plagued with problems of erratic supply and low quality. Thailand is the largest cassava starch exporter. Its modern starch mills produce a steady supply of a high quality product.

The international starch market is highly complex, and with the rapid development of modified starches, developed countries may be difficult markets to enter. Cassava starch is, however, preferred for the preparation of certain foods and for papermaking and the cardboard industry, indicating a continued but probably small export demand.

Starch is an important input in the textile industry, where it is used both as a lubricant and for sizing. In developing countries that have large textile industries, the potential demand for cassava starch may be substantial. The possibilities of using cassava starch as a sweetener have not been examined in great detail. In developed countries, fructose syrups are being used increasingly to replace sugar, and since the raw material in the production of fructose syrup is starch, cassava might be used as the basis for a fructose syrup industry. Indonesia, which imported 782,000 tons of sugar in 1981, has built several plants to produce fructose syrup in order to replace imports.

Alcohol

In the 1970s, when the Brazilian national petroleum company, Petrobras, made plans for replacing much of the gasoline it uses with alcohol, there was considerable skepticism. Increasing prices of petroleum products have, however, shown that the Brazilian approach to the energy crisis may have considerable potential. As the supply of petroleum products becomes tighter, demand for petroleum substitutes such as alcohol will increase, and it will be difficult to satisfy this demand. Government agencies in Colombia, Thailand, the Philippines, Zambia, Papua New Guinea, and Zimbabwe are studying the feasibility of producing cassava alcohol as a petroleum substitute. Cassava is a particularly attractive source for alcohol production since it can be produced on marginal land and need not compete for land used for food crops.

In 1980, Brazil produced enough alcohol to replace 20 percent of the gasoline needed for automobiles in that country. Most of the alcohol was obtained from the fermentation of sugarcane, which was possible because of the low world price of sugar and a surplus sugar production in Brazil. However, in the future, it will not be possible to obtain increased alcohol supplies from sugarcane, which requires prime agricultural land, without impairing the country's capacity to produce food. Consequently, Brazil plans to greatly expand the production of alcohol from cassava grown in less-favored agricultural areas.

A major question concerning the production of alcohol from cassava is the energy balance of the system. When sugarcane is used to make alcohol, the bagasse is burned to provide energy. No technology for using cassava stalks in a comparable manner has yet been developed, though the use of the stalks would greatly improve the net energy ratio. In Brazil, where the cassava distilleries burn firewood, the total system of cassava production, forestry, and alcohol production was analyzed, and the net energy ratio was found to be close to 9 to 1. However, other calculations are not so favorable—some are as low as 1.2 to 1.

The energy used in the distillation process is considerable (see Table 11), but technological advances should reduce the distillation requirements considerably. Alcohol-tolerant yeasts, for example, can increase the ethanol (alcohol) content of the fermentation liquor from 6 to 12 percent and reduce the energy required for distillation by 30 percent. Furthermore, most of the estimates for cassava are based on rather conservative yield figures relative to the level of agricultural inputs, such as fertilizers, used in the calculations. As new production technology raises crop yields, the net energy ratio of cassava alcohol should improve. Even now the net energy ratio is positive, although at times it is relatively small. Development of improved technology for conversion to alcohol should also enhance the net energy ratio.

Producing alcohol from cassava cannot be recommended indiscriminately as a solution to the energy problems of developing nations. Alcohol production will surely always take second place in the allocation of resources after food production. As a result, cassava alcohol production will only be feasible in countries that have large areas of land that are not needed for food production and that have a cassava production potential that is greater than the direct demand for food. Furthermore, with present technology, distilleries require a cheap energy source for fuel to make the process worthwhile.

The Brazilian cassava project has caused concern that the consumption of cassava by the distilleries might cause shortages of the roots and thus drive up the price of cassava for human consumption. This problem seems unlikely because the success of the alcohol project depends on a cheap and abundant supply of cassava. There has also been concern that the new alcohol plants may attract large growers who will swamp the small growers. The Brazilian government has attempted to avoid this problem by constructing "mini-factories" (*mini-usinas*), which can buy from the small farmer. The presence of nearby mini-factories should stabilize cassava prices in an area and give farmers a larger and more dependable market.

The development of a national alcohol program will create new opportunities for cassava growers and lead to greater employment and wealth in the rural areas. Roads, electricity, and other forms of infrastructure will be required if the project is

Table 11
Net energy ratio (NER) of cassava and sugarcane alcohol (produced by various methods)

Raw material	System		Energy produced (kcal/ha)	Energy consumed (kcal/ha)			NER[a]
	Fuel	Drying		Agricultural transportation	Distillery	Total	
Sugarcane			5,590,000	640,000	20,000	660,000	8.47
Cassava	Cassava stalks	Sun	5,570,000	700,000	420,000	1,120,000	4.97
Cassava	Cassava stalks	Mechanical	5,570,000	760,000	680,000	1,440,000	3.87
Cassava	Oil		5,570,000	480,000	4,110,000	4,590,000	1.21

Source: T. P. Phillips, Economic implications of new techniques in cassava harvesting and processing, in *Cassava harvesting and processing: Proceedings of a workshop held at CIAT, Cali, Colombia 24–28 April 1978,* ed. E. Weber, J. H. Cock, and A. Chouinard, 66–74 (Ottawa, Ont.: IDRC, 1978).

[a] Per cubic meter of absolute alcohol.

to be successful, and those improvements should directly benefit the rural population. In addition, most of the equipment needed for the alcohol plants is made in Brazil, and its construction and later maintenance and operation are done by Brazilians. Thus the alcohol program will stimulate industrial development and employment within Brazil.

3
Cassava production

Most of the world's cassava is grown by small farmers who employ traditional production systems, which give low but stable yields. Areas that produce cassava tend to have few roads or other infrastructure, and the soils generally are poor. Growers use very labor-intensive agronomic practices; only land preparation is fairly often mechanized.

FARM SIZE AND LABOR USE

In Asia, farms of cassava growers are small. Even in Thailand, where cassava is grown almost exclusively for export, the average cassava producer has only 3.3 hectares. In India, where cassava is mainly produced for local consumption, cassava growers' farms are usually less than 1 hectare, and there are many backyard patches in urban areas. Indonesia has a few large cassava plantations, but most of the farms are very small. In both the Philippines and Malaysia, cassava is grown as a backyard crop and as a plantation crop. Plantations may have 20 hectares or more of cassava, but their aggregate output is only a small fraction of total production.

Information about the average farm size of cassava producers in Africa is scanty. In southern Nigeria, where most of the country's cassava is grown, three-fourths of the farms are under 1 hectare. The lack of mechanization and the preponderance of shifting cultivation in Africa suggest that that farm size may be quite typical of the continent.

Farms tend to be larger in South America. Of the area planted to cassava in the three primary producing countries—Brazil,

Paraguay, and Colombia—40 to 50 percent is on farms of 10 hectares or less. There is tremendous variability in farm size, even between regions within the same country. One survey in Ecuador found that two-thirds of the cassava producers plant fewer than 6 hectares of cassava, but farm size ranged from less than 1 hectare to over 900 hectares. In one of the zones studied, the majority of the farms were smaller than 2 hectares, while in another zone, the majority of the farms were larger than 10 hectares. Despite the size of their farms, South American cassava growers are not necessarily more prosperous than farmers elsewhere. Much of the land used for cassava has extemely low fertility and is considered marginal for agricultural production. The majority of the cassava producers have very limited resources to work with.

Most cassava is produced by hand except that land preparation is mechanized in some countries. In Colombia, a survey showed that slightly more than half of the farmers plow with tractors, and the proportion is rising. In Thailand and Malaysia, too, most of the land for cassava is prepared mechanically. In India and Indonesia, however, much of the tillage is done manually or with draft animals. In Africa, land for cassava is usually tilled by hand because the crop is grown predominantly in areas that lack draft animals due to the presence of sleeping sickness spread by the tsetse fly. Manual land preparation is slow. In one area of Colombia, land clearing and preparation required 33 man-days per hectare, and throughout Colombia, it averaged 25 days. Similar amounts of time are spent in Zaire (see Table 12). There are considerable differences in the time required for the various operations. In shifting cultivation in Zaire, planting is the most time-consuming operation, probably because there is little prior working of the soil. But because the farmers clear the ground by burning off vegetation, weeds are suppressed, and as a result, weed control requires less labor in Zaire than in Colombia or Thailand. In the latter two countries, weed control is almost entirely manual and is the most laborious activity in producing the crop.

Harvesting and packing takes considerable labor, too—up to 50 man-days per hectare in Zaire but only 20 man-days per hectare in Thailand, where yields are slightly greater. The time

Table 12
Labor use for producing cassava in three countries (worker-days/ha)

	Colombia		Thailand	Zaire	
Task	A	B	A	BC	BD
Land preparation		25	10	25	16
Planting	9	11	10	35	35
Weeding	47	44	33	10	10
Harvesting and packing	31	25	20	50	50
Total	87	105	73	120	111
Yield (t/ha)	11	11	15	10	10
Worker-days/ton	8	10	5	12	11

Sources: Colombia: R. O. Diaz, P. Pinstrup-Andersen, and R. D. Estrada, *Costs and use of inputs in cassava production in Colombia: A brief description* (Cali, Colombia: CIAT, 1975); Thailand: T. P. Phillips, A profile of Thai cassava production practices, in *Proceedings of the Fourth Symposium of the International Society for Tropical Root Crops held at CIAT, Cali, Colombia, 1–7 August 1976,* ed. J. H. Cock, R. MacIntyre, and M. Graham, 228–232 (Ottawa, Canada: IDRC, 1977); Zaire: B. F. Johnson, *The staple food economies of western tropical Africa* (Stanford, Calif.: Food Research Institute, Stanford University, 1958).

Note: A = land prepared mechanically; B = land prepared manually; C = after clearing forest; D = after clearing forest regrowth

needed for harvesting and packing is greatly influenced by the ease of harvesting the variety, the type of soil, and the way the roots are packed.

The total labor requirements in the three studies shown in Table 12 are rather similar when allowance is made for mechanical or manual preparation of the land. Other studies, however, have found much higher labor use elsewhere: 200 man-days per hectare in Nigeria; 300 man-days per hectare in Indonesia, and as an average for all of Africa; and 100 man-days per hectare in the Philippines for harvesting only. Little information is available to explain these extraordinarily high figures. In general, however, it seems that with mechanical land preparation, 1 hectare of cassava requires 70 to 90 man-days of labor and with manual land preparation, it requires 100 to 200 man-days.

Per hectare, cassava requires more labor than most other starchy staples. However, when the comparison is made in terms of labor input per calorie harvested in traditional agricultural systems, cassava's labor requirements are not excessive.

PRODUCTION AREAS

Although cassava is grown under diverse conditions throughout the tropics, several broad categories encompass the vast majority of cassava production areas.

Lowland tropical areas with a pronounced dry season

Nearly half the land planted to cassava is in lowland tropical areas that have a pronounced dry season. Examples are northeastern Brazil, the north coast of Colombia, India, Thailand, Java, and parts of Zaire and Tanzania. These areas receive 800 to 2000 millimeters of rain a year and have a 3- to 6-month dry period in which less than 10 percent of the total annual rainfall occurs. Mean annual temperatures are greater than 22° C. Soils are usually light in texture, of inherently low fertility, and acid, with pH values reaching as low as 4.5.

Hot, humid lowland tropics

Cassava growing areas in southern West Africa, parts of Zaire, the Amazon jungle, and the less densely populated areas of Indonesia are classified as hot, humid lowland tropics. Over 1500 millimeters of rain falls annually; the dry season (or seasons) is normally shorter than 3 months. The mean temperature is 22° C or greater with little seasonal variation. The natural vegetation, tropical rain forest or regrowth, has to be cleared before planting.

The acid-infertile savannas

There are vast areas of underutilized acid-infertile soils in the tropics, most of which are covered by jungle or savanna vegetation. These soils, called Oxisols and Ultisols, occupy some 1600 million hectares in 48 developing countries. The savanna areas have tremendous potential for crop production of species that are well adapted to these conditions. The climate of savanna areas is characterized by a pronounced 3- to 6-month dry season. Soils have pH values as low as 4.0 and often contain levels of exchangeable aluminum high enough to be toxic to crops. In Colombia, Venezuela, Brazil, and Bolivia, savannas are traditional cassava growing areas.

Areas with a cool winter season

Cassava is grown as far from the equator as 30° north and 30° south latitudes. These areas have a cool winter—frosts occur in the more extreme situations, and snow occasionally falls—but the annual average temperature is usually over 20° C, and summer temperatures are always high. Annual rainfall is 1000 millimeters or more, with most of the rain falling during the hot summer months. Cassava is grown under these conditions in southern Brazil, Paraguay, northern Argentina, Cuba, southeastern United States (Florida), Taiwan, and southern China.

Highland tropics

Cassava production at elevations of 1500 to 2300 meters above sea level occurs only in the Andes of Peru, Ecuador, and Colombia and the mountainous regions of tropical Africa. Annual rainfall in these highland areas is over 1500 millimeters, but it is rather uniformly distributed throughout the year. The average annual temperature is between 17° C and 22° C.

PRODUCTION PRACTICES

The broad array of ecological and sociological conditions under which cassava is grown gives rise to extreme diversity in production systems and practices. Nevertheless, certain production practices are widely used. Annual yields of more than 70 t/ha of fresh roots have been obtained on experimental plots, and yields of over 40 t/ha a year have been reported by many experimental stations in different parts of the world. In Colombia, some farmers have obtained yields of over 40 t/ha, and in some years the average yield of good farmers in the Caicedonia area has been estimated at 35 t/ha.

Among countries with more than 50,000 hectares planted, the highest yields are in Angola (14.2 t/ha) and India (16.7 t/ha). World average yields, however, are less than 9 t/ha, and close to a quarter of the countries that grow cassava have yields of less than 4 t/ha. Thus even the highest national yields are well below those obtained on experimental plots, and yields in most countries are a mere fraction of the potential.

A few studies have been made of the causes of low yields in farmers' fields in relation to those obtained on experiment stations and in trials established in farmers' fields. Studies in Thailand, Nigeria, Colombia, and Ecuador have highlighted the importance of pests and diseases, soil fertility, mixed cropping, poor quality of planting material, suboptimal agronomic practices, and lack of high yielding, disease- and pest-resistant varieties in keeping farmers' cassava yields at a low level.

Land preparation

Cassava is grown in shifting culture in much of Africa and parts of Asia and the Americas. Farmers cut forest or savanna vegetation and burn it before hoeing the land. They then dig holes, place organic matter in them, and heap up the soil to produce mounds on which cassava and other crops are planted. In parts of Colombia where erosion is a problem, farmers do not prepare whole fields for cassava; rather, they till small patches (about 50 x 50 centimeters) on which the cassava is to be planted.

When the entire field is tilled, it may be done by hand, with draft animals, or by tractor. When preparation is by mechanical means, the land is normally plowed and than harrowed or disked. In areas where drainage is a problem, the land is heaped in mounds or ridges, and the cassava is planted on the crest.

Varieties

Cassava production has not had the benefit of the decades of research that have been devoted to cereal crops. Most cassava varieties are traditional clones that have been selected by farmers. Since cassava is an introduced species in Africa and Asia, farmers have made their selections from a limited genetic pool. Due to a lack of variability within *Manihot esculenta*, scientists in Africa,

India, and Indonesia have turned to interspecific crosses as a means of increasing variability.

In Indonesia, Mukibat, a farmer, developed a grafting technique, which now bears his name, in which "tree cassava" (*Manihot glaziovii*) is used as the scion and cassava as the root stock. This system and others based on multiple grafts are used by many small farmers in eastern Java. Although it is doubtful that grafting is advantageous in large plantings, it allows extremely high yields to be obtained from a small number of plants grown around a house. The vigorous top growth also provides shade for the living quarters.

A small number of clones dominate Asian production areas: in India, M-4 introduced from Malaysia; in Malaysia, Black twig and Green twig; in Thailand, Rayong 1. In West Africa and the Congo basin, there is a wider range of genetic variability, and farmers often plant several different clones in the same field. Latin America has much greater variability in its clones than do the other continents. Within the regions of Latin America, there are large numbers of local clones, each of which has special characteristics. Farmers often grow several clones and keep naturally occurring hybrids that germinate from sexual seed. These collections of cassava are tested and observed by the farmers and may eventually be used in their cropping areas.

The potential for yield increase by varietal improvement is enormous. In trials carried out by Centro Internacional de Agricultura Tropical (CIAT) under a wide range of ecological conditions, the best lines outyielded the local lines by more than 50 percent, even with minimal levels of fertilizers and fungicides. Similarly high yields of new hybrids have been obtained by the Central Tuber Crops Research Institute in India and the International Institute of Tropical Agriculture in Nigeria.

New varieties are not yet widely distributed, and this factor also contributes to the yield gap. Care should be taken, however, before massive multiplication and distribution of new varieties are attempted. Experience gained in both India and Colombia, where the fresh market is important, has highlighted the importance of quality to consumers. High yielding varieties may not be more profitable for farmers unless those varieties are well accepted. When lines have poor root quality, high yield may be

Because it spoils quickly, freshly harvested cassava must be taken to market every day. (*Source:* Ulrich Scholz)

inadequate to compensate for the low price they fetch in the market.

In addition, many clones do not adapt easily to new conditions. For example, in Colombia, a clone selected at a lowland site was, without further testing, recommended for planting in a nearby region at a different altitude. The new, high yielding clone failed when it succumbed to a disease that was not present in the warmer temperatures of the initial testing site.

These words of caution are offered to explain why high yielding varieties are not in common use rather than to belittle the possibility of their becoming widely adopted. In southern Brazil, the hybrid variety Mantiqueira, developed by the Instituto Agronomico de Campinas, is grown over a broad area, and yields are high. This variety is also grown commercially in Cuba, the Dominican Republic, and Colombia.

Planting material

The planting material for cassava is the stems of mature plants (at least 8 months old). At harvest time, farmers set aside stems as planting material for the succeeding crop. The material may be planted immediately or stored for up to 6 months if the harvest period does not coincide with planting time. Planting material is taken from the woody part of the stems. If it is intended for storage, it is cut into stakes at least 1 meter long and kept in the shade, or it is placed in bundles with the basal parts buried in the soil in the shade of a tree. The stored stakes produce shoots from their upper ends, and the shoots are trimmed off before planting. In southern Brazil, where cuttings have to be held through the cool winter, they are stored in underground shelters to protect them from frost damage.

Farmers have difficulty in assessing the quality of planting material. They pay attention to the germination of the planting material and initial vigor of the new plants, but they usually do not know the nutritional and phytosanitary status of the original cuttings. Even though different planting materials may have similar germination or sprouting percentages, and though all new plants may appear equally healthy, yields may differ greatly depending on the nutrition and disease status of the original plants from which the cuttings were taken. The keys to choosing planting material that will give satisfactory yields are careful selection of cuttings, treatment, and proper storage (see Chapter 4).

Cuttings for the next planting are taken from mature plants just before harvest. (*Source:* CIAT)

Planting: Timing and method

Just before planting, the planting material is cut into pieces 10 to 30 centimeters long, but occasionally longer. These cuttings are then planted horizontally, vertically, or inclined. In horizontal planting, the pieces are buried 5 to 10 centimeters below the soil surface. In vertical or inclined planting, only a half to two-thirds of the length of the pieces is covered with soil. If ridges or mounds have been prepared, the stakes are planted on the upper part. In parts of Cuba, where water is applied in furrows after planting, the cuttings are planted at the bottom of the ridges, but this planting practice is disappearing because it results in low yields and root rots.

From 7000 to 20,000 stem pieces are planted per hectare. High populations are used if the soil fertility is low, if cassava is being grown in monoculture, or if erect, low-branching varieties are being planted. The most common population is about 10,000 plants per hectare.

Although cassava is relatively drought tolerant once established, it requires adequate soil moisture for the first 2 to 3 months after planting. In areas with a dry period, cassava is normally planted at the beginning of the wet season, but secondary plantings are often made 2 to 3 months before the onset of the dry season.

Soil fertility

Cassava is usually grown without fertilizer on soils that have inherently low fertility. In addition, it is frequently the last crop in the rotation before a fallow period because it is able to grow and yield well, relative to other crops, in poor soils. Farmers are well aware of the ability of cassava to grow on depleted soils. The Campa Indians of Peru, for example, plant cassava as a sole crop on the steepest, least-fertile slopes in their slash and burn agriculture. On more level, more fertile land, they intercrop cassava with maize in the first planting, and as fertility declines in subsequent plantings, they lower the density of maize plants and increase the density of cassava plants.

Farmers who use no fertilizer must leave the land fallow periodically. In a fallow or shifting culture, the yield of cassava is closely related to the duration of the fallow and the number of crops that have preceded the cassava since the last fallow. But population pressure in many countries has become too great to allow large tracts of land to lie fallow (or to revert to forest or savanna vegetation) long enough to regain fertility. Consequently, cassava yields decline.

In a few countries, small farmers who are short of land and are unable to maintain long fallows are beginning to apply fertilizer to cassava. On large plantations in Malaysia, cassava has been continuously cropped for up to 15 years, and yields have been maintained, in the absence of disease problems, if chemical fertilizer has been applied.

Some growers who use no chemical fertilizer on cassava apply fertilizer to the cash crops that precede the cassava in the rotation, and the cassava exploits the residual fertilizer. This practice is common in parts of Brazil and Peru where cassava follows tobacco.

Cassava has long been thought to "deplete" or "exhaust" the soil. For this reason, the British Colonial Service, early in this century, attempted to restrict cassava growing in Malaya, markedly reducing the cassava area there. More recently, the Thai Ministry of Agriculture became concerned about criticisms of cassava and attempted, unsuccessfully, to limit cassava growing in that country.

Many trials have shown that cassava does extract significant amounts of nutrients from the soil. Per unit of dry matter produced, however, the quantity of nutrients, other than potassium, that it takes from the field is no greater than that extracted by many other crops (see Table 13). This fact suggests that cassava's reputation for depleting the soil is undeserved. It probably comes from cassava's ability to grow on soils already so depleted that most other crops cannot be grown on them. In such circumstances, it is not due to cassava per se that other crops cannot be grown on those soils once a cassava crop has been grown and harvested. Yet it is impossible to expect cassava to reach its yield potential on depleted soils unless fertilizer is applied. The lack of even moderate fertilizer use in cassava production is a major factor in keeping farmers' yields far below the potential.

On experiment stations, the response obtained from applying fertilizer is often greater than that encountered on farmers' fields because scientists normally grow their trials under optimum conditions. Fertilizer application may increase disease incidence and weed competition, hence it is important for researchers to avoid viewing the application of fertilizers in isolation. Rather, it should be considered as part of the improved technology that can be used to reduce the yield difference. In some regions in

Table 13
Soil nutrient extraction by various crops

Crop	Nutrient extraction (kg/ton of dry matter harvested) N	P	K	Ca	Mg
Cassava (roots)	6	1	11	1.6	0.6
Barley (grain)	21	5	6	0.9	1.4
Maize (grain and cob)	15	3	6	0.5	1.7
Potatoes (tubers)	10	2	22	0.8	1.3
Rice (grain with hulls)	13	3	4	0.4	1.6
Sorghum (grain)	20	4	4	0.5	1.9
Wheat (grain)	18	4	6	0.6	1.8

Sources: J. A. Nijholt, *Absorption of nutrients from the soil by a cassava crop* (in Dutch), Buitenzorg Algemeen Proefstation voor den Landbouw, Korte Mededeelingen no. 15 (Buitenzorg [Bogor], Indonesia, 1935); National Academy of Sciences, *Nutrient requirements of beef cattle* (Washington, D.C., 1970); National Academy of Sciences, *United States–Canadian tables of feed composition* (Washington, D.C., 1969).

which cassava is intensively cultivated, such as Cuba, southern Brazil, and Australia, fertilization is common and no doubt contributes to the high yields.

Mixed cropping

Interplanting cassava with another crop generally reduces yields. In trials in Colombia, for example, intercropping reduced average yields by an estimated 1.9 t/ha; in India, where cassava is frequently grown under coconut trees, yields are reduced greatly.

The "gap" between farmers' yields and experiment station yields is in part due to mixed cropping. Although 30 to 40 percent of the world's cassava is grown in association with other crops, almost all yield data from experiment stations refer to yields in monoculture. But since the total production of food per hectare may well be increased by intercropping, the resulting reduction in cassava yield should not be viewed with alarm.

Cassava fits intercropping systems well, particularly with grain legumes. (*Source:* CIAT)

Weed control

Weed control is important in the early growth stages because a cassava crop is slow to establish and cover the ground. Weeding is often the most labor-intensive aspect of cassava production. Land preparation provides initial weed control, but later, up to six or seven weedings with a hoe or a machete may be necessary. Once the crop is well established, shade from the cassava leaf canopy prevents weeds from offering strong competition.

Pests and diseases

Cassava is generally thought to be extremely tolerant of disease and pest attack. This perception stems largely from its tolerance to attack by locusts in Africa; but a large number of experiments have shown that other pests and diseases rarely cause complete crop failure, though they may severely depress yields. Surprisingly, the yield losses are not greatest in the most conspicuous attacks by pests, such as hornworms, that sporadically cause complete defoliation. Rather, the most costly diseases and insects are those that are present throughout the crop's growth cycle, such as African mosaic disease, cassava bacterial blight, Cercospora leaf spots, and spidermites.

Losses vary tremendously from region to region. Latin America, the center of origin of cassava, has all major problems except locusts and African mosaic disease. Historically, Africa has been in closer contact with the Americas than with Asia, mostly because of the slave trade. This contact led not only to an exchange of cassava clones, but also to a transfer of the diseases and pests that attack them. Asia on the other hand has been relatively isolated from Latin America, and its disease and pest incidence is much lower than Africa's (see Table 14).

There is little information about crop losses that result from diseases and pests in farmers' fields. Estimates from a survey in Colombia indicated that superelongation disease and Cercospora leaf spots, when present, each reduced yields by 3 t/ha or more. In Nigeria, yield losses due to cassava bacterial blight have been estimated to be as high as 75 percent, and in Zaire in the mid-1970s, an epidemic of the disease caused severe yield reduction. In another instance, the introduction of cuttings infected with cassava bacterial blight lowered yields by approximately 50 percent on a large plantation in Brazil.

In Africa and India, where African mosaic disease is endemic, yield reductions are estimated to be as high as 90 percent, but they certainly average much less. This disease is spread by whiteflies, but man is also a prominent vector when infected cuttings are used. There is some evidence that African mosaic disease could be eradicated from cassava growing areas by the use of clean planting material; however, with the exception of India and Zanzibar, no schemes exist for producing mosaic-free cuttings for farmers.

In the Americas, where cassava was domesticated, natural predators and parasites of cassava pests have evolved and exist

Table 14
Distribution of major pests and diseases of cassava

	Latin America	Africa	Asia
Pests			
Thrips	•	•	
Mites	•	•	•
Hornworm	•		
Fruitfly	•		
Shootfly	•		
Whiteflies	•	•	•
Stem borers	•	•	
White grubs	•	•	•
Cutworms	•		
Gall midges	•		
Lace bugs	•		
Grasshoppers		•	
Mealybugs	•	•	
Scales	•	•	•
Leaf-cutter ants	•	•	
Crickets	•	•	
Termites	•	•	•
Diseases			
Cassava bacterial blight	•	•	•
Superelongation disease	•		
Frogskin disease	•		
Phoma spp.	•	?	?
Cercospora spp.	•	•	•
Anthracnose	•	•	
Erwinia	•		
African mosaic		•	(India only)
Root rots	•	•	•
Pathogens of planting material	•	•	•

Sources: A. Bellotti and A. van Schoonhoven, *Cassava pests and their control* (Cali, Colombia: CIAT, 1978), and J. C. Lozano and R. H. Booth, Diseases of cassava, *PANS* 20 (1974):30–54.

in a delicate balance with the pests. In addition, farmers have selected lines that are tolerant to many pests, and they often plant several different lines or clones together to lower the risk of pest damage. The balance is well maintained in traditional farming systems in which pesticides are seldom used. With an indiscriminate use of pesticides, however, losses can be dramatically increased.

When new pests or diseases occur in Africa and Asia, they may do far greater damage than they do in their center of origin because natural predators and parasites often do not exist and because no farmers have selected tolerant lines. For example, in Africa, outbreaks of the green spidermite and mealybugs cause severe losses. These two pests apparently were introduced to Africa relatively recently, and with few natural enemies being present there, populations built up rapidly, leading to heavy damage. In the Americas, both of these pests are problems, but not on the same scale as in Africa.

The combined effect of different diseases and pests on the yield of cassava is not easily measured. However, diseases and pests certainly are a major reason yields are lower in farmers' fields than at experiment stations. Use of resistant varieties, clean planting material, and integrated pest management could substantially reduce yield losses.

Harvesting

Unlike crops such as rice and wheat, cassava can be harvested more or less whenever it is needed. The harvest may begin as soon as 7 months after planting in warm areas, or it may be put off until 18 months after planting, or later. Long growth periods are common in areas that have cool winters and in highland areas where average temperatures are low. Since the roots' starch content tends to be greatest when temperatures are low, cassava is commonly harvested during the cooler months. When the rains begin after a long dry season, the starch content drops sharply, and harvesting tapers off.

To prepare cassava for harvest, most farmers cut the tops, leaving a stump about 30 centimeters long, which is grasped and used to uproot the plant. If the soil is hard, farmers may tie a pole to the stump to use as a lever. Mechanical harvesters and other harvesting aids are in limited use.

4
Components of a new technology

Cassava has been regarded as a rugged, hardy crop that is tolerant of disease and pest problems and is grown mostly by subsistence farmers. In the past, that reputation has kept countries from giving cassava serious attention, but Brazil and India are important exceptions. In Brazil, during the 1940s and 1950s, the Instituto Agronomico de Campinas mounted a sustained cassava research program. The improved agronomic practices and varieties that resulted account for the high average yields, 19 t/ha, in the state of Sao Paulo, where the institute is located. In India, during the last 25 years, the Central Tuber Crops Research Institute has been developing agronomic practices that, along with the introduction of the variety M4 from Malaysia, have increased yields in the state of Kerala from 5 t/ha to more than 15 t/ha. Elsewhere, significant research on cassava was done by the Dutch in Indonesia before World War II, the French in Madagascar, and the British in Kenya. These early research efforts have formed the basis for the development of new, improved technology.

In 1967, de Vries, Ferwerda, and Flach published their now-classic paper on the potential of different food crops in the tropics. They concluded that the root and tuber crops, particularly cassava, had an extraordinary potential for producing calories (see Table 15). At about the same time, a group of scientists, mainly from the West Indies, formed the International Tropical Root and Tuber Crop Society, which, through triennial symposia, has stimulated scientific interest in these crops.

In 1971, work on cassava began at two international agriculture research centers. At the International Institute of Tropical Agri-

Table 15
Maximum recorded yield (in dry weight and food energy) of tropical food staples

Crop	Maximum annual yield (t/ha)	Food energy per day (kcal/ha)
Cassava	71	250
Maize	20	200
Sweet potato	65	180
Rice	26	176
Sorghum	13	114
Wheat	12	110
Banana	39	80

Source: C. A. de Vries, J. D. Ferwerda, and M. Flach, Choice of food crops in relation to actual and potential production in the tropics, *Neth. J. Agr. Sci.* 15 (1967):241–248.

culture in Nigeria, the tropical root crops improvement program concentrates on cassava for African conditions, while at the Centro Internacional de Agricultura Tropical in Colombia, the cassava program develops production systems for the Americas and Asia. National cassava programs have been established in many developing countries during the past two decades. In addition, various agencies in developed countries have contributed to solving problems associated with cassava production and utilization. A new technology has emerged, not as the result of any breakthrough, but from assembling information obtained by many researchers over many years and combining it in production packages. The exact composition of each production package will differ from others, depending on local conditions, and hence, adaptive research will be required as the packages are introduced into new areas.

The new technology is evolving rapidly. The worldwide research effort is expanding as interest in the development of the crop rises on the part of national and international agencies. Certain components of the improved technology are now well proven; however others are still being tested. The latter is particularly true for new varieties because breeding, selection, and testing require considerable time. Nevertheless, many components of the new technology are already available, and others are in advanced stages of development and show every indication that they will shortly have a favorable impact on meeting the goal of increased productivity.

IMPROVED VARIETIES

Most cassava varieties are the outcome of selection by generations of farmers, so they are adapted to the ecological conditions of the locality in which they are grown. Particularly in Latin America, where cassava originated, there is an enormous array of varieties that thrive in the localities in which they were selected. But when these varieties are moved away from their home areas, they often fail completely.

Only a few varieties yield well under diverse ecological conditions. An example is M Col 1684, a clone collected from Leticia in the Amazon jungle, which has shown extraordinary yield potential over a wide range of conditions in several countries (see Table 16). Although the variety performs poorly in cool climates and in areas where thrips are a problem, it has demonstrated the possibility of obtaining varieties that can yield well in several environments.

The Centro Internacional de Agricultura Tropical (CIAT) and the International Institute of Tropical Agriculture (IITA) are making superior germ plasm and varieties available to national scientists for possible use in their breeding, selection, and testing programs. So far no "miracle" cassava variety has been found that would result in a dramatic increase in yield potential comparable to the one offered by modern short-statured wheat and rice varieties. However, cassava germ plasm with high yield potential, selected from Latin American germ plasm, is now being incorporated into breeding programs in the Americas and Asia. This same germ plasm has been crossed with disease-resistant lines at IITA and distributed throughout Africa.

It appears unlikely that any one variety will be found that will serve all cassava growing areas. Rather, varieties must be selected for the specific conditions under which they will be

Table 16
Yield of the clone M Col 1684 under a wide range of conditions

Site	Site features			Yield (t/ha)	
	Soil	Length of dry period	Average temp.	M Col 1684	Best local variety
Caicedonia, Colombia	fertile	short	22°C	57.1	45.2
Carimagua, Colombia	acid infertile	medium	26°C	36.1	10.1
Media Luna, Colombia	infertile	long	27°C	40.7	8.8
Santa Clara, Costa Rica	moderately fertile	short	22°C	35.0	19.7
Pichilingue, Ecuador	fertile	short	25°C	42.4	18.0
San Cristobal, Dominican Republic	moderately fertile	short	25°C	39.9	24.6

Source: CIAT, *Annual report 1978* (Cali, Colombia, 1979).

cultivated. CIAT and IITA follow a similar decentralized selection process to obtain improved varieties. This breeding strategy involves producing elite gene pools for each of the major soil/climate zones in which cassava is grown. Germ plasm accessions and promising hybrids are planted in the major zones, and clones that show useful characters—such as disease and pest resistance, high starch content, and good yield potential—are then crossed at headquarters. The progeny of these crosses are returned for further selection to the original sites or distributed to national breeding programs that are breeding for similar conditions. In this manner, CIAT and IITA make large numbers of sexual seeds or smaller numbers of clones available to national programs so they can select varieties that are well adapted to specific local soil and climatic conditions.

Cassava breeding is a slow process. It takes at least 10 years from the time a cross is made to the release of a fully tested variety. At present, there is a tremendous amount of promising material that has high yield potential as well as disease and pest resistance, that should be ready for release in the coming years.

Yield potential

Even after the roots start to fill, the cassava crop continues to produce leaves, which convert solar energy into chemical energy in the form of carbohydrates. Consequently, for maximum root growth, there must be a delicate balance between the production of leaves and their supporting stems on the one hand and the growth of roots on the other. The plant must have enough leaf area to produce carbohydrates, but it should not produce so much leaf and stem that no carbohydrate remains available to fill the roots. This balance is shown schematically in Figure 4 (leaf area index is the ratio of the total area of the leaves to total ground area). Maximum root growth, it is worth noting, occurs at leaf area indices that are considerably lower than those required for maximum production of total dry matter, that is, biomass.

This insight into the physiological basis of yield has allowed the formulation of a plant ideotype for maximum yield. The

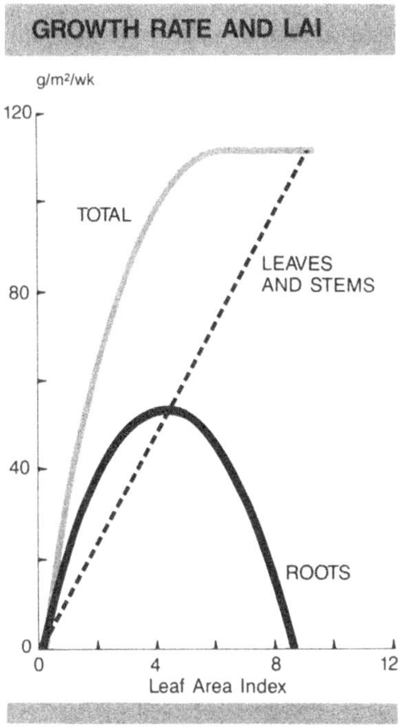

Figure 4. Physiological explanation of the optimal leafiness for maximum yield.

ideal plant type would rapidly reach a leaf area index of approximately 3 and maintain that index throughout the rest of its growth cycle. The cassava plant requires large leaves in order to build up leaf area rapidly and a long individual leaf life and late branching in order to maintain its leaf area index at 3 in the later growth states (see Table 17).

The easiest parameter for the breeder to measure on individual plants that integrates most of these factors is the harvest index—root weight divided by total plant weight. The yield of an isolated plant is not correlated with its yield when it is grown as part of a plant community. The harvest index of a plant, however, has a loose but useful relationship to the plant's yield in a community. Maximum yields are obtained by plant types that have harvest indexes ranging from 0.5 to 0.7.

A computer model that describes the growth of such an ideal plant type under good conditions with moderate levels of solar

Table 17
Characteristics of an ideal cassava variety for use under good growing conditions, planted at 10,000 plants/ha

Branching:	Late branching at 6 and 9 months after planting. No side or sucker branches.
Leaf size:	Large maximum leaf size (ca. 500 cm/leaf) 4 months after planting.
Leaf longevity:	Long leaf life—approximately 100 days.
Harvest index:	Harvest index greater than 0.5.
Leaf area index:	Between 2.5 and 3.5 for most of the growing cycle.
Roots:	Nine or more roots per plant when planted at 10,000 plants/ha.
Shoot number:	Each plant should have two main shoots from the original cutting.

Source: J. H. Cock, O. Franklin, G. Sandoval, and P. Juri, The ideal cassava plant for maximum yield, *Crop Science* 19 (1979):271–279.

radiation suggests that the potential annual yield of cassava is 90 t/ha of fresh roots, or 30 t/ha of dry roots. On small plots at CIAT in Colombia, a dry root yield of 28 t/ha per year has been obtained from a hybrid line that has been bred to closely resemble the ideal plant type described above, and many hybrids yield more than 20 t/ha of dry roots per year. Thus, under good conditions, a yield potential of 30 t/ha of dry roots per year appears reasonable.

The use of the harvest index has enabled breeders to identify very high yielding clones, but the first clones selected in this manner have not been planted by farmers to any great extent because of their low quality, which is usually related to low starch content or high cyanide levels. Nevertheless, these clones have demonstrated that selection for high yield can be achieved, and new clones without these deficiencies are now in the advanced testing stage.

To develop varieties with high yield potential for varied conditions, the CIAT program chooses materials for crossing at sites that are representative of the major cassava growing areas.

Then, at the CIAT headquarters' station, where flowering is profuse, crosses are made specifically to produce varieties for each of the major growing areas, and the materials are sent to the various sites for further selection. The selection sites are the north coast of Colombia, representative of the lowland tropics with a pronounced dry season; the eastern plains of Colombia, representative of the acid-infertile savannas; the tropical rain forest in lowland Colombia east of the Andes; the highland areas of Colombia; and Cuba, where there are marked seasonal changes in temperature and day length. In the selection process, fungicides, pesticides, and irrigation are not used, and lime and fertilizers are applied, at low levels, only in the acid-infertile savannas. Yields vary greatly from site to site: up to 50 t/ha in the more fertile areas of the north coast and up to 20 or 30 t/ha in the eastern plains where soils are infertile and disease and pest pressures are high.

Disease and pest resistance

Because cassava is a long-season crop that is grown as a source of cheap calories, it is not economical for the farmer to use heavy applications of fungicides and pesticides against diseases and pests. Hence breeders pay a great deal of attention to incorporating pest and disease resistance into new lines of cassava. The fact that a cassava crop is normally grown throughout the year, with no break in the cycle that would lower populations of disease pathogens and pests, has had evolutionary consequences. Cassava in the Americas, its center of origin, has evolved as a crop that has high levels of resistance to the pathogens and pests that are present in the environment in which it is grown. Furthermore, the type of resistance that has evolved is generally of the horizontal, or field, type rather than vertical, or single-gene, resistance, which frequently breaks down as new pathotypes occur. Horizontal resistance, while not conferring immunity, does not break down. Breeders of crops that rely on vertical resistance are continually forced to search for new resistance genes and to incorporate them when previously used resistance sources become ineffective.

Cassava bacterial blight, probably the crop's most costly disease, is widely distributed throughout the Americas, Asia, and Africa. Typical symptoms are gray, angular leaf spots that develop into large brown areas with a water-soaked appearance. When the infection is acute, the stems and petioles exude a gummy substance, and the shoots may die back. The disease is particularly severe when rainfall is high and there are large fluctuations in daily temperature. It is mainly spread by splashing water in the rainy season and the use of infected cuttings. Many sources of resistance to the disease have been found, and a major part of the breeding work at IITA and CIAT is the production of lines that are resistant to cassava bacterial blight. Both institutes have found that resistant lines maintain their resistance when moved to other areas. Furthermore, plants grown from true seeds produced by crosses between resistant parents have shown a high frequency of resistance.

African mosaic disease causes heavy yield losses in Africa and India. Somewhat surprisingly, it is not found in the Americas where cassava originated. Characterized by a mosaic pattern on the leaves and leaf distortion, the disease is thought to be caused by one or more viruses and is disseminated by whiteflies, infected cuttings, and possibly by contaminated machetes used in preparing planting material. Crosses between cassava and tree cassava, *Manihot glaziovii*, that were made in Kenya many years ago produced the Amani lines, which have high levels of resistance. They are being widely used in breeding programs in Africa and India. IITA has used them and new crosses to produce a large number of lines that are resistant to both cassava bacterial blight and African mosiac disease. These lines have been distributed in the form of pure seed throughout Africa and India, and resistant clones that are suitable for various ecological conditions are being selected.

Superelongation disease was first identified in the 1970s in the Americas, where it is now known to be widespread. It can cause severe losses in crops that are attacked early in the season. The symptoms are a marked elongation of the internodes and a malformation of the leaves. It is spread by infected cuttings and spores that may travel long distances, but infection occurs only under humid conditions. Resistant varieties have been identified at CIAT, and efforts are being made to combine resistance to cassava bacterial blight and superelongation disease, as these are the two most serious diseases in the Americas.

The most destructive insect pests of cassava are spidermites and mealybugs. Spidermites attack mainly in the dry season, and the symptoms vary depending on the species of mite involved. The green spidermite (*Mononychellus tanajoa*) was introduced to East Africa in the 1960s from the Western Hemisphere. In the absence of any natural enemies, and with large areas planted to susceptible varieties, the insect has spread rapidly and is now causing extensive damage from Tanzania in East Africa to the Guinea coast in West Africa. In East Africa and South America, differences in varietal resistance have already been encountered, but it will take time for breeders to incorporate resistance into high yielding lines.

Mealybugs do severe damage in Africa and northeastern Brazil, where they have few natural enemies and most local clones are highly susceptible to the pest. The predominant species are different. *Phenacoccus manihoti*, which is found in southern South America, was introduced to Africa relatively recently. *P. herreni* is native to the Americas, but it appears to have been introduced into northeastern Brazil in recent years. Although few resistant lines have been identified so far, levels of resistance probably can be raised by breeding programs.

Thrips are a widespread pest in the Americas. They lower yields by distorting and reducing the size of new leaves, stunting the new growth of the plant, and damaging the apical meristem, which may be killed in severe attacks. Like the spidermite, thrips are common during dry periods. Many resistant clones have been found, and as a result, most breeding programs are producing resistant lines for areas in which thrips are a problem.

Many other diseases and pests attack cassava, but they are of lesser importance or are significant only in certain localities. Phoma leaf spot, for example, is the most serious disease of cassava in highland equatorial areas. High levels of resistance exist, and special lines are being selected for this rather limited cassava growing area.

Numerous diseases and pests attack cassava, but its long growth cycle still allows it to tolerate sporadic destruction of its

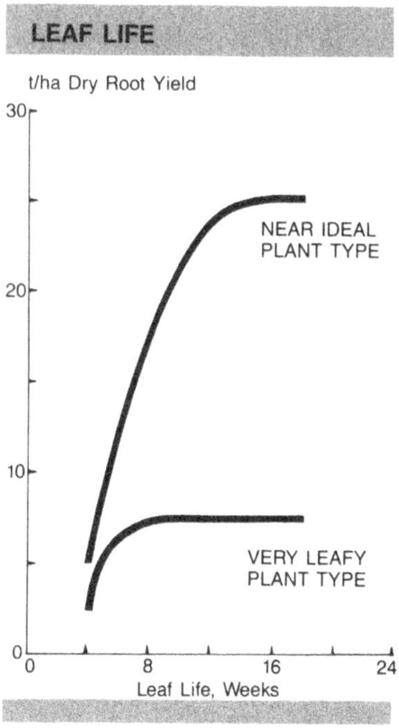

Figure 5. Simulated yields of a near-ideal and a traditional vigorous cassava variety as related to leaf life (pest or disease attack shortens leaf life).

leaves—one complete defoliation lowers yields by about 20 percent. However, pests or diseases that are permanently present may depress yields severely. By using a growth-simulation model, the traditional leafy cassava type has been compared with the near-ideal plant type in terms of the effect of a continued attack of a disease, such as Cercospora leaf spots, that shortens leaf life (see Figure 5). For the traditional leafy type, with its low yield potential, reduced leaf life has little impact on yield. The near-ideal plant type is much more sensitive to reduced leaf life (and hence disease), though its yield potential is greater at all levels of disease damage. This fact suggests that although cassava is generally hardy, disease and pest control are of greater importance for yield stability in the new varieties that have high yield potential than it is in the traditional types.

Root quality

The criteria for root quality vary considerably from region to region. In Latin America, consumers in some areas prefer cassava roots with a white skin, while others, often within the same country, prefer roots that have a brown skin with a pink undersurface. Such color preferences usually have arisen because they are characteristic of a particular local clone that has good eating quality. It is probable that the introduction of new high-quality lines with a different skin color will be accepted. In one area in which the preferred local variety had a rose-colored underskin, a local experiment station began a series of trials with a white-skinned type of good eating quality. Theft from the station became a severe problem, and the white-skinned cassava began to be available in the local markets. At first, it sold at a discount, but as the quality of the white-skinned cassava became recognized, it sold at a premium.

Good root quality tends to be associated with low hydrogen cyanide content, a high starch content, and long shelf life. It is widely believed that bitter cassava—cassava with high levels of cyanide—yields more than the sweet types. Although this may once have been true, in the new lines being produced there is no evidence that the sweet types have a lower yield potential. Although no cassava types without cyanide have been found, clones that contain very low cyanide levels are now being developed through selection.

Lines with a high starch content and other desirable characters such as disease resistance and high yield potential are now in the advanced testing stage in many countries. Unfortunately, high starch varieties tend to have poor storage properties. It is unwise for cassava programs to concentrate on breeding for starch (or dry matter) content while paying insufficient attention to solving storage problems.

At various times, breeders have attempted to produce high-protein cassava varieties, and expectations have been raised by reports of success. Results of experiments with high-protein lines must be interpreted carefully. The protein content of cassava roots is often high before root filling occurs, hence varieties that are harvested early or that are inherently low yielding tend to have high protein levels. There is good reason to believe that selecting for high protein content must eventually lead to a lower yield potential because the synthesis of protein requires almost twice as much of the primary products of photosynthesis as the synthesis of a similar weight of starch.

Varietal release

Small cassava farmers in traditional cassava growing areas often plant two or more cassava varieties in the same field, and they constantly try new materials. In the Amazon region of Colombia, farmers observe the performance of plants that grow from true seed, and they select the ones that have desirable characters and multiply them. These farmers are, in effect, running their own breeding programs; however the recombinations are from uncontrolled crosses, and the numbers involved are very small, which limits the farmers' chances of obtaining superior new lines. Breeding programs can bring together a much larger gene pool and can make crosses that are directed toward combining several different desirable characters in one line. By testing large numbers, the probability of a successful development of improved varieties is greatly increased.

Once a breeding program has identified particular lines as promising, they are tested to see if they maintain yield and quality characteristics from generation to generation and also to see if they are adaptive to different growing conditions. In most crops, these ends are achieved by establishing a large number of replicated yield trials, but in traditional cassava growing areas, promising materials may be distributed to farmers even before they have been tested sufficiently for formal release. The farmers will grow and test them and select the best ones for multiplication. The information gained from the farmers will help the breeders decide which varieties to release, and the risks to the farmer are minimal.

In crops that have a rapid propagation rate, it is dangerous to give farmers, or seed companies, materials that have not been tested, because the crop may become widely grown before a serious weakness becomes apparent, possibly leading to a disaster for the farmers. With cassava, however, the propagation rate is so slow (unless special techniques are adopted) that before a variety is widely adopted, any severe shortcomings will be recognized and farmers will be cautious about expanding their plantings of the variety. This system also has the advantage that the farmers will multiply materials so that by the time the variety is officially released, sources of cuttings will be available. In short, in traditional cassava growing areas, the risks of a new variety becoming widely grown and then failing are minimal. If a variety is seriously flawed, farmers will reject it long before its failure causes widespread problems.

The situation is quite different, however, when cassava is introduced into a new area for large-scale production. When any new crop is introduced, or the area planted is rapidly increased, the crop tends to be relatively free of disease and pest problems for a few years. But with time and as the area planted expands, problems begin to appear. In high-value crops, chemicals may be utilized to control diseases and insects, but with a low-value crop such as cassava this course of action is impractical. Thus care must be taken to obtain cassava varieties that are tolerant to the pests and diseases that are likely to occur as the crop area increases. This aim can best be achieved by carefully monitoring pest and disease incidence in new varieties; by selecting varieties that are known to be tolerant of the disease and pest complex in areas with similar climate and soils; and by recommending several varieties, differing in genetic background, rather than just one, so that if one line fails, others will be available. In addition, rapid propagation facilities should be established so that as problems emerge, new tolerant lines can be rapidly multiplied to replace the susceptible ones.

Rapid propagation

In the field, cassava normally can be multiplied 3- to 30-fold per year, depending on the variety and the growing conditions,

In the high humidity of propagation frames, cuttings rapidly produce multiple shoots. (*Source:* CIAT)

a rate of multiplication that is much slower than that of most major crops, particularly ones that are reproduced sexually. In recent years, however, two simple techniques for rapid propagation have been developed. They allow more than 1 million stakes or cuttings to be produced from one original cutting in less than 2 years. These techniques are highly labor intensive, so the chief limit to the rapid propagation of cassava is now economic rather than biological.

The simplest method is based on the facts that very young cassava shoots can be rooted in water and that axillary buds will sprout once apical dominance is broken. Two-node cuttings of mature cassava plants are planted in special propagating frames that maintain high temperature and humidity. The cuttings sprout, and when the shoots are about 10 centimeters long, they are cut off near the base—leaving the lowest axillary buds still attached to the original cutting—and rooted in water. Once the cuttings have produced roots, they can be transplanted directly

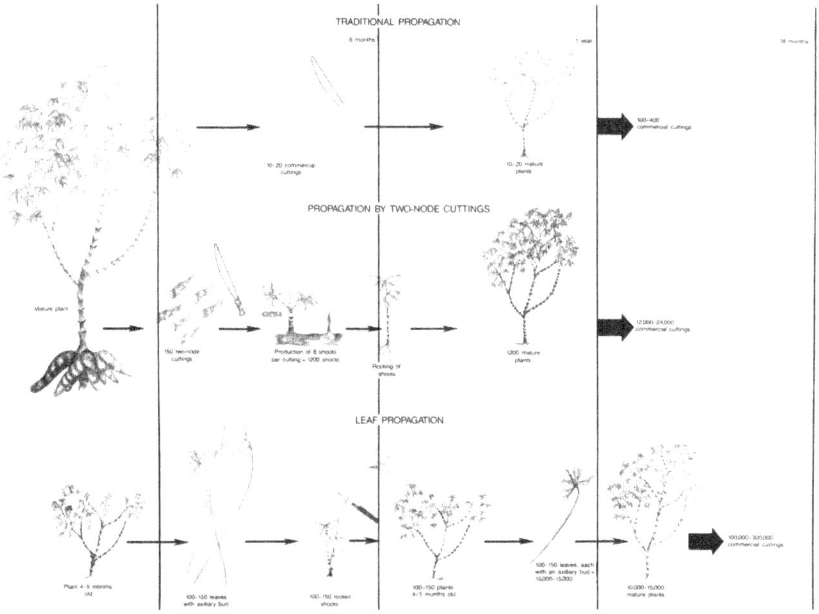

Figure 6. Three propagation methods: traditional, two-node cuttings, and leaf propagation.

to the field. The axillary buds on the original cutting sprout again, and the process can be repeated several times. On average, each two-node cutting can be expected to produce 8 to 12 shoots for rooting. The process is shown schematically in Figure 6.

The second propagation method requires more care and has not been very successful in areas that have a cool winter period, though in other areas, it has proved reliable and rapid. It involves excising propagules from 3-month-old cassava plants grown in the field. A propagule is a leaf and its petiole plus the bud at the base of the petiole and a small heel of stem tissue. The leaf lamina are cut in half (see Figure 7) to reduce transpiration, and the propagule is placed in a misted chamber. After about 2 weeks, the propagule sprouts and produces roots. It can then be transferred to a peat pot, and after 2 or 3 weeks, it can be transplanted in the field. Three months later, the whole process can be repeated.

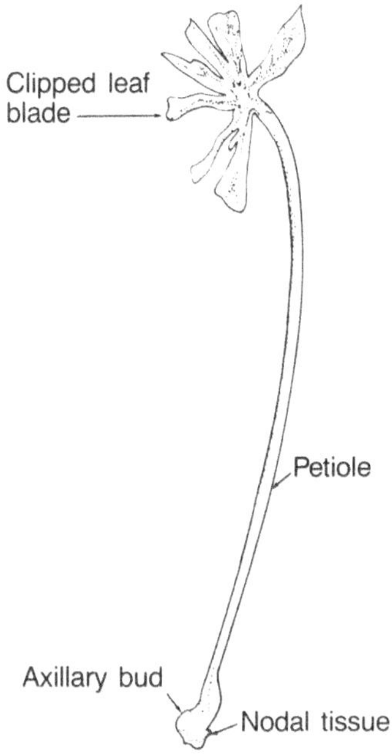

Figure 7. A propagule, which is central to the leaf propagation system.

FERTILIZER REQUIREMENTS AND SOIL AMENDMENTS

On low-fertility soils with little or no fertilizer—conditions under which few crops will grow—cassava is capable of producing modest yields. Cassava does, however, respond to fertilizer applications, and in many areas, yields can be sharply increased by fertilization.

It is important to determine the fertilizer requirements of cassava experimentally in different soils. Most crops show well-defined symptoms of deficiencies of major nutrients (nitrogen-deficient cereal crops, for example, have yellow leaves), but cassava displays symptoms only in extreme cases. Usually the cassava plant adjusts its top growth in relation to the availability of major nutrients. In this way, the plant maintains a high nutrient status in the leaves, which appear normal and healthy. Thus it is difficult to visually assess the nutrient status of a cassava crop—a healthy-looking crop may have a severely limited yield potential because of a lack of nutrients.

For maximum root yield, the appropriate fertilizer level is well below the level required for maximum plant growth. Heavy applications of fertilizer, particularly nitrogen, can stimulate top growth to such an extent that although total plant weight increases, the yield of roots may decrease. There is no evidence to suggest that the tendency for cassava yields to decrease if high levels of fertilizer are applied will disappear with the development of high yielding, lodging-resistant varieties, unlike the experience with modern wheat and rice varieties. Consequently, for maximum root yield as opposed to total plant weight, cassava has relatively low nutrient requirements.

Making fertilizer recommendations for farmers who use cropping systems involving fallows of varying lengths can be perplexing. Generally, as population density increases, pressure on land becomes greater and fallow periods are shortened, which results in a decline in soil fertility. Nevertheless, when cassava is planted after a fallow period, it frequently produces high yields and does not give an economic response to the application of fertilizer. However, as successive crops are grown without fertilizer, yields decline, and after several cycles, yields cannot be raised to the immediate post-fallow levels by applying fertilizer. On the other hand, if small annual applications are made, yields can be maintained at close to the immediate post-fallow level. The problem for extension agents is that they should recommend fertilizer applications in the first year after fallow, even though there is no economic response, in order to maintain yields over several years.

Nitrogen

Since rising agricultural productivity throughout the world has been closely associated with an increased use of nitrogen fertilizer, it is perhaps surprising to find that cassava requires relatively little nitrogen to achieve high yields. Other than in peaty soils, cassava rarely responds to more than 100 kg/ha of nitrogen, and yields often decline markedly as nitrogen levels are increased.

Phosphorus

In the vast areas of acid-infertile soils, which have a tremendous potential for cassava production, phosphorus deficiency is a major limitation. On these soils, cassava has a yield response to applications of as much as 400 kg/ha of phosphorus, and levels of 100 to 150 kg/ha are frequently recommended.

Such heavy fertilization can be done cheaply with rock phosphates, according to studies by the International Fertilizer Development Center in cooperation with CIAT. Ordinarily, the phosphorus in rock phosphate is not in a form that is readily available to plants, but in acid soils, in which cassava grows satisfactorily, its availability increases. Treating rock phosphate with an acid solution (partial acidulation) before application to increase the availability of phosphorus results in cassava yield responses that are comparable to those obtainable from the application of triple superphosphate, but at a fraction of the cost. Since initial applications of rock phosphate have a long-term residual effect, the amount of later applications can be reduced.

Experiments conducted in nutrient solution show that cassava has a higher phosphorus requirement than most crops, but in the field, this high requirement is often not evident because cassava roots normally harbor mycorrhizae, which increase phosphorus absorption. On degraded soils, in which native mycorrhizal populations may be very low or in which the strains present may be ineffective, yields may be depressed. Some preliminary tests suggest that inoculation of planting material with effective strains can increase yields; however practical field methods of producing the inoculant have yet to be developed.

Potassium

Cassava extracts more potassium from the soil than any other element—a good crop extracts 100 kg/ha or more. To maintain yields when cassava is grown continuously, potassium fertilization is essential. Potassium fertilization also affects quality. If potassium is deficient, the dry matter and starch content of the roots falls, and the cyanide content rises. The application of potassium is particularly important on sandy soils and acid-infertile soils.

Other elements

Little attention has been paid to sulfur fertilization of cassava, but because the crop is generally grown on poor soils far from industrial centers that release sulfur into the air (which is returned to the soil in the form of acid rain), sulfur deficiency can be expected. A marked response to sulfur has been obtained in the eastern plains of Colombia. Sulfur can be applied by using potassium or calcium sources in the form of sulfates.

Calcium is normally applied to cassava in the form of lime as an amendment to raise soil pH. The plant's response to lime may often be more the result of a correction of calcium deficiency than from raising soil pH, because cassava is extremely tolerant of low soil solution pH. Cassava rarely responds to the application of more than 2 t/ha CaO, even on strongly acid-infertile soils. Most of the yield responses occur from applications ranging from 0.5 to 1 t/ha. There is some evidence, however, that higher lime application levels may be necessary to achieve maximum yields when heavy applications of nitrogen, phosphorus, or potassium are made. It is preferable to apply dolomitic limestone because magnesium deficiency is also common in acid soils.

Zinc deficiency frequently occurs in cassava fields and is aggravated by heavy lime applications. Unlike deficiencies of major nutrients, which usually do not produce clear symptoms in cassava, those of the minor elements, particularly zinc, are distinctive. Leaves of zinc-deficient plants show yellow or white mottling, often in a chevron pattern. The deficiency can easily be corrected by dipping planting material in a dilute solution of zinc sulfate, or through foliar applications of zinc sulfate.

Cassava as a crop tolerates the high levels of aluminum and low pH that are characteristic of many tropical soils. If the aluminum level is such that the cation-exchange capacity is less than 80 percent saturated, no adverse effects occur. When aluminum levels are greater, lime can be applied to reduce the aluminum saturation to an acceptable level, which results in large yield increases.

In summary, cassava is rather tolerant of low fertility, soil acidity, and high levels of aluminum. It does, however, respond well to fertilizer application and has a particularly high requirement for phosphorus and a rather low nitrogen requirement. Potassium fertilization is necessary to maintain soil fertility and is important for obtaining good yields of high quality cassava. When cassava is grown continuously, annual applications of the major nutrients are essential to maintain soil fertility and yields. On acid soils, positive responses to lime can be obtained, but care must be taken not to overlime, which could induce minor-element deficiencies. Zinc deficiency, the most common minor-element problem, can easily be remedied by treating the stakes before planting.

WEED CONTROL

Since newly planted cassava grows slowly, it is vulnerable to weed competition. For high cassava yields, weeds must be controlled for the first 3 to 4 months after planting, depending upon growing conditions, until the crop completely covers the ground. Without weed control, it is impossible to obtain good yields of cassava.

Hand weeding

Cassava needs two to six hand weedings depending on the severity of weed competition and the time it takes the crop to cover the ground. Hand weeding is effective and can be used when labor is both plentiful and cheap. If weeding is delayed because labor is scarce, yield can be greatly reduced.

Herbicides

There are many commercial chemicals that can adequately control weeds in cassava fields for the first 2 months after planting. Diuron works well against annual weeds and alachlor against grasses. A mixture of the two applied before cassava emerges is particularly effective. Beyond 2 months after planting, this mixture has to be supplemented with hand weeding or a directed application of a contact herbicide.

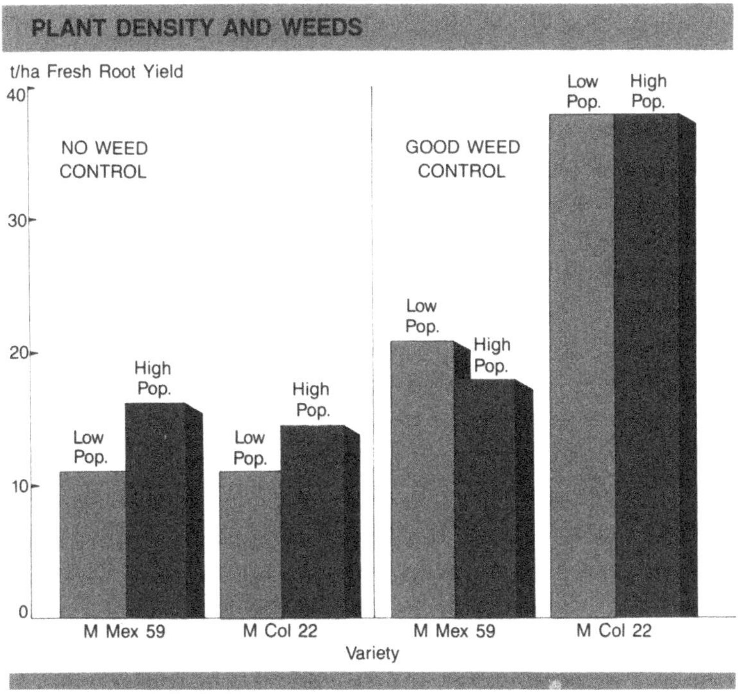

Figure 8. In the absence of weed control, high planting population helps cassava compete with weeds. (*Source:* CIAT *Annual report 1979*)

A new preemergence mixture, oxifluorfen mixed with alachlor, may eliminate the need for any hand weeding. This mixture also has some value against purple nutsedge (*Cyperus rotundus*), one of the most difficult weeds to control.

Cultural control

Vigorous cassava varieties, which rapidly cover the ground, tend to suppress weeds and require less weed control. Unfortunately, vigorous varieties usually are too leafy and have a low yield potential, so even though they may yield well with poor weed control, their yields will not equal those of less vigorous types under good management. By planting less vigorous varieties densely, rapid ground cover can be achieved, thus improving the crop's ability to compete with weeds (see Figure 8).

The timing of planting can influence the amount of weed control that is needed. If cassava is planted at the beginning of the rainy season when weeds flourish, the farmer has to struggle to weed the cassava as well as the other crops he has planted. If cassava is planted somewhat before the beginning of a short dry season, when farm labor demands are lighter, the crop can be weeded once or twice, or sprayed with a preemergence herbicide mixture, until the dry season begins. At the onset of the dry season, weeds cease to be a serious problem. When rainy weather resumes, the cassava will rapidly cover the ground before weeds can germinate and compete.

Another way to control weeds during cassava's early growth is to interplant short-statured grain legumes (see the section "Mixed cropping" later in this chapter). Because the cassava crop does not intercept much light during its early growth, the grain legume is not shaded excessively so the grain legume grows rapidly and smothers weeds that would compete with the cassava. The legume crop can be harvested about the time the cassava crop is starting to cover the ground. Thus good yields of both grain legumes and cassava can be obtained with minimal weed control (see Figure 9).

Integrated weed control

A farmer is likely to use a combination of weed control methods rather than just one. Intercropping holds back weeds, but some active control usually is still necessary for optimum yields. Hand weeding is often the least expensive system, but when weeds are growing vigorously, there may be labor shortages and consequently, the farmer may turn to herbicides because timeliness is essential for good weed control.

INTEGRATED PEST MANAGEMENT

As a result of being continually exposed to pathogens and pests during its long growth period, not only has cassava evolved a generally high level of resistance to local diseases and insects, but a large number of predators, parasites, and pathogens of cassava pests have evolved as well. The cassava plant, its diseases

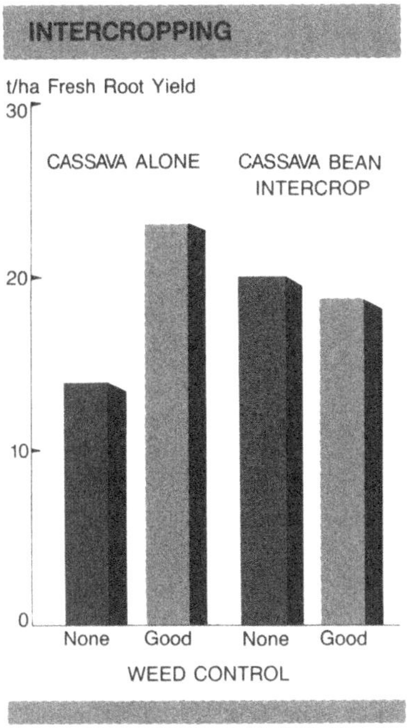

Figure 9. Intercropping with grain legumes reduces the amount of weed control needed for good cassava yields. (*Source:* CIAT *Annual report 1978*)

and pests, and the factors that control those diseases and pests maintain a delicate, but dynamic, ecological balance. The aim of integrated pest management systems is to favor cassava without destroying this balance.

An integrated pest management system employs manifold control measures that are compatible with one another to reduce diseases and pests to a level at which damage to the crop is slight. This reduction does not necessarily entail completely eliminating a disease or pest but, rather, holding damage to the crop to an acceptable level. Cassava researchers have emphasized the use of stable host-plant resistance, biological control, cultural and phytosanitary practices, and, as a last resort, a limited use of chemicals. The major diseases and pests of cassava are listed in Table 18, along with suggested control methods.

Table 18
Major diseases and pests of cassava and control measures

Disease or pest	Losses	Distribution	Control methods
Cassava bacterial blight	to 100%	Widespread	Clean seed, agronomic practices, resistant varieties[a]
African mosaic disease	to 90%	Widespread, Africa, India	Clean seed, roguing, resistant varieties
Superelongation	to 100%	Limited, Americas	Stake treatment, varietal resistance[a]
Frogskin disease	to 100%	Very limited, Americas	Disinfection of tools,[a] clean planting material[a]
Phoma	to 100%	Cool humid areas	Varietal resistance[a]
Cercospora leaf spots	to 30%	Very widespread	Varietal resistance[a]
Pathogens of planting piece	to 100%	Very widespread	Stake treatment
Anthracnose	Not known, may be high	Limited, Americas and Africa	Varietal resistance[a]
Preharvest root rots	to 100%	Mainly poorly drained areas	Crop rotation, ridging

Green spidermites	to 50%	Widespread, in dry season	Biological control,[a] varietal resistance[a]
Hornworm	20% per attack	Widespread in Americas only	Biological control
Thrips	to 30%	Widespread	Varietal resistance
Scales	Reduces germination, up to 20% from later attacks	Widespread	Biological control,[a] treatment of planting material[a]
Mealybugs	Probably high	Limited	Biological control,[a] varietal resistance[a]
Shootflies	Very low	Widespread in Americas	Only necessary in early growth stages, varietal resistance[a]
Whiteflies	to 80%	Americas	Varietal resistance[a]

Sources: A. Bellotti and A. van Schoonhoven. *Cassava pests and their control* (Cali, Colombia: CIAT, 1978), and J. C. Lozano and R. H. Booth, Diseases of cassava, *PANS* 20 (1974):30–54.

[a] Methods being developed or most likely to be developed.

Varietal resistance

In developing varietal resistance, breeders are concentrating on horizontal, or field, resistance, which does not confer immunity on the plants but does limit the multiplication of the pests or pathogens as well as the damage they cause. Losses can further be reduced in many cases by combining varietal resistance with preventive measures such as the use of "clean" planting material.

For small farmers with limited resources, varietal resistance is a particularly suitable control method. The major expenses can be borne by the research and development agencies that produce the new varieties. The only cost to farmers is the purchase of a small quantity of foundation "seed," which they can use to produce new planting stock and, simultaneously, roots for sale or consumption. Furthermore, apart from the initial multiplication and distribution of new varieties, the introduction of resistant varieties places little burden on agricultural service agencies. There is no need, for example, to launch training programs or to develop new credit facilities. Although research and development programs on new varieties may appear costly, the returns on investment can be extremely high.

Varietal resistance is not a panacea, however, as not all diseases and pests can be controlled in this way. Furthermore, because the rate of progress in a breeding program is generally inversely proportional to the number of breeding objectives, a program that attempts to solve all disease and pest problems by breeding may move so slowly that it reaches no worthwhile goal within a reasonable amount of time. Hence, in practice, breeding for varietal resistance is restricted to the most serious disease and pest problems encountered.

Biological control

Biological control is being used against three important cassava insects, the cassava hornworm, green spidermites, and mealybugs. The cassava hornworm (*Erinnyis ello*) is a caterpillar that develops into a migratory moth. No source of varietal resistance to this voracious pest has been found. Normally, populations of horn-

Farmers in Colombia maintain a population of Polistes wasps, predator of cassava hornworm, by placing the wasps' nest in palm-thatched huts near cassava plantings. (*Source:* CIAT)

worms are held at low levels by predators and parasites, but occasionally, attacks of hornworms defoliate whole cassava growing areas, with each defoliation causing yield losses of around 20 percent. When a severe hornworm outbreak occurs, the farmer's first reaction often is to apply a potent, broad-spectrum insecticide, which kills not only the hornworm but all of its natural enemies as well. Subsequently, in the absence of predators and parasites, hornworm populations expand quickly, and another attack occurs, necessitating further insecticide applications. Apart from the pernicious environmental effects of repeated heavy applications of insecticides, this method of control is expensive for the farmers and destroys many beneficial insects that control other cassava pests.

Research at CIAT has shown that hornworms can be controlled by introducing two wasps, *Polistes* spp. and *Trichogramma* spp., into cassava fields and spraying a bacteria in extreme cases. Farmers in the Caicedonia region of Colombia are using this biological control system because of a program organized by the

National Coffee Federation. This form of control has reduced hornworm outbreaks and decreased the severity of damage at low cost. Under the program, cassava growers maintain a population of Polistes wasps, which attack hornworm larvae, by placing the wasps' nests in the cassava fields in small structures covered with palm fronds. This control measure is enhanced by the use of battery-operated light traps that attract hornworm moths. If the light traps indicate that the adult moth populations are high and if a large number of hornworm eggs are observed on the leaves of the cassava plants, farmers buy Trichogramma wasps, which are commercially available, and release them to parasitize the eggs.

Occasionally, despite these measures, massive migrations of the adult moth occur, and larval populations build up. Then farmers apply *Bacillus thuringiensis* to hold the damage to low levels. This bacteria infects the hornworm without affecting the natural enemies of cassava pests. A disadvantage of this treatment is that even though the hornworms cease feeding within hours after *B. thuringiensis* is applied, they take several days to die. Farmers like to see dead hornworms, and it is difficult to convince them that sick hornworms do little or no damage. Hence the effectiveness of this control measure requires the support of a strong extension program.

Although sources of resistance to green spidermites are being incorporated into breeding programs, the levels of resistance are not high. Particularly in Africa, great emphasis is placed on supplementing resistance with biological control methods. IITA is coordinating a massive effort to rear and release natural enemies of the green spidermite throughout cassava growing areas in Africa. Phytoseiidae, a predator of mites, shows particular promise. Since it appears to be effective even when spidermite populations are low, it could prevent explosive outbreaks. This project may greatly reduce the damage these insects cause.

The cassava mealybug *Phenacoccus manihoti* was only recently discovered in its center of origin in southern South America. IITA and CIAT are collaborating on the collection and evaluation of natural enemies of the cassava mealybug for eventual mass rearing and release in Africa.

Misgivings about the introduction of natural enemies of both the green spidermite and the cassava mealybug from the Americas to Africa are sometimes expressed on the supposition that these insects may themselves become pests in Africa. But experience with using biological control agents with other crops indicates that this problem will not arise. On the other hand, when the natural enemies are introduced into Africa, they should be completely free of their own natural enemies, particularly parasites, that could cripple their effectiveness.

Naturally occurring populations of predators and parasites keep populations of other potentially serious cassava pests at low levels. In the Americas, scales and mealybugs are seldom damaging because they have numerous natural enemies. Hence growers should avoid applying pesticides indiscriminately, as this action could suppress the beneficial insects.

Clean planting material

The two most important cassava diseases, African mosaic disease and cassava bacterial blight, are disseminated by infected cuttings. By planting disease-free cuttings, farmers can delay the onset of severe infection until late in the growing season. In some instances, the use of clean planting material may even eliminate a disease.

In Kenya and India, mosaic-free planting material has been obtained through the selection of symptomless plants from infected fields followed by careful roguing. Both of these countries have found that disease-free plantations only slowly become reinfected. In India, the Central Tuber Crops Research Institute is multiplying mosaic-free seed stocks for release to farmers; a similar program might be successful in Kenya. It is still unclear whether the experience in India and Kenya can be repeated elsewhere; however, in areas with a low incidence of whiteflies (the vector of the disease), the use of mosaic-free planting stocks could probably greatly reduce the incidence of the disease or even eliminate it from plantations.

It is rather simple to produce material that is free of cassava bacterial blight. Cuttings with no obvious symptoms of the disease are allowed to form shoots, which are cut off when they reach about 8 centimeters in length and are rooted in small flasks filled with water. Any plantlets that are infected develop disease symptoms rapidly and can be destroyed. When the symptom-free plantlets have produced roots, they are transplanted into an isolated field where they form the basis for disease-free stock. This technique has been used in Colombia, Brazil, Cuba, and Malaysia to produce blight-free planting material. The Instituto Colombiano Agropecuario uses disease-free foundation stocks to multiply blight-free material for sale to farmers. In Costa Rica and the Caicedonia area of Colombia, where farmers have planted only blight-free material in recent years, the disease has been eradicated. Attempts to eliminate cassava bacterial blight in areas where environmental conditions are extremely favorable to the disease have been less successful. Even in those areas, however, use of clean "seed" will delay infection enough to moderate yield losses considerably (see Figure 10).

Superelongation, another disease that is disseminated by infected cuttings, can be eliminated from cuttings by treating them with fungicides before planting, because the pathogen is present only in the superficial layers of the planting material.

Cultural and phytosanitary practices

If cassava is planted continuously or in a very short rotation in high-rainfall areas, wet root rots caused by organisms such as *Phytophera* spp. may cause severe losses. The only remedy is to leave the soil fallow or to rotate it with other crops, preferably cereals, that do not act as an alternative host for the pathogen. Improving soil drainage retards the buildup of the disease, but the expense is rarely justified for cassava production. Planting on ridges, however, can improve drainage sufficiently to prevent heavy damage from wet root rots. Ridging is advantageous in areas in which the annual rainfall exceeds 1500 millimeters. In West Africa, farmers plant on mounds, which similarly improves drainage and reduces root rots.

Cassava is susceptible to attacks by certain pathogens of forest species. In Malaysia, white-thread disease (caused by *Phomes lignosus*) often attacks cassava that is planted after jungle or

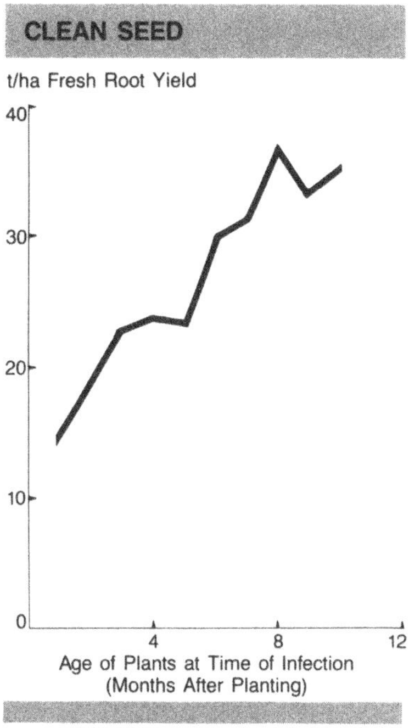

Figure 10. By delaying the onset of cassava bacterial blight infection, use of clean "seed" can increase yields. (*Source:* CIAT *Annual report 1976*)

rubber trees have been cleared. In areas in which this disease is severe, it is prudent to plant another crop before introducing cassava.

Rainfall pattern influences the incidence of diseases and pests. When rainfall is heavy, diseases such as cassava bacterial blight, superelongation, anthracnose, and Cercospora leaf spots increase. In the dry season, pests such as thrips, mealybugs, and spidermites predominate. In certain areas in which cassava bacterial blight is severe and local varieties are only moderately resistant, the disease can greatly depress yields when the crop is planted at the beginning of the rainy season. If, however, the crop is planted toward the end of the wet season, the onset of dry weather retards the buildup of bacteria. Thus, in some areas, cassava bacterial blight can be partially controlled by shifting

In high-rainfall areas, planting cuttings vertically on high ridges combats root rots. (*Source:* CIAT)

the planting period. For example, in trials in the eastern plains of Colombia, yields have been almost doubled by planting at the end of the wet season rather than, as is usual, at the beginning of the rainy season. At present, cassava is a subsistence crop in the region, and the small, isolated plots farmers plant at the start of the rainy season are less severely affected by diseases than large commercial fields would be. But if the cassava area is to be expanded with moderately resistant varieties, the planting date should be changed to achieve better disease control.

CULTURAL PRACTICES

Agronomic practices have been developed by farmers over thousands of years, and the production systems they use are influenced by the conditions under which the crop is produced. Nevertheless, there are a large number of methods to improve cassava production that are applicable to a broad range of production conditions.

Planting material

Good quality planting material is crucial for high yields, yet most farmers use poor planting material because the typical crop from which they take cuttings is diseased, is damaged by insects, or has been inadequately fertilized. Furthermore, if planting does not immediately follow the harvest, the vegetative planting pieces must be stored, during which time their capacity to germinate declines, as does their ability to support vigorous early growth. A substantial buildup of insect pests such as scales on stored planting material may occur as well.

The cuttings selected for planting should have at least half their total diameter occupied by pith and be at least three nodes long. Farmers should discard cuttings that show signs of mechanical or insect damage on the surface or that have a discolored central pith. In one trial in Colombia, plants grown from cuttings that had been damaged by a fruit fly and a bacterial pathogen yielded a third less than plants grown from cuttings that had no visible damage. In areas in which soil fertility is low and disease and pest pressures are high, it is advisable to reserve a small portion of a cassava field for planting material and to give this area extra fertilizer and added protection from diseases and pests.

Cuttings should be dipped in fungicide and insecticide before planting. Several suitable dips for cuttings are shown in Table 19. After being soaked in the solution for 3 minutes, the stakes should be allowed to dry before being stored or used for planting. This inexpensive treatment protects planting pieces from attack

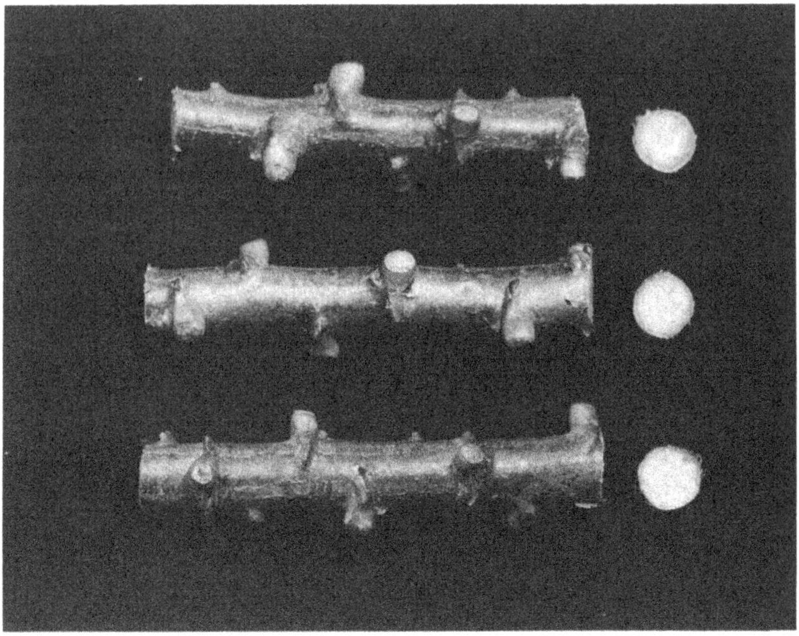

Good quality cuttings for planting. (*Source:* CIAT)

by soil-borne pathogens and by surface pests such as mites and mealybugs, thus greatly improving the germination percentage. The yield increases will more than cover the cost.

When cuttings are stored (sometimes for as long as 5 months), they sprout, which lowers their carbohydrate and nutrient reserves. The cuttings may also become dehydrated and infested with various pathogens. If selection is rigorous, a considerable quantity of cuttings will be lost before planting. A simple remedy is to immerse the long cassava stems in a fungicide-insecticide mixture both before storage and again just before planting. On farmers' fields in a poor area of Colombia, a combination of stake treatment and selection was a major factor in increasing yields of the local unimproved variety by 70 percent.

"Seed banks"

An alternative to storing cuttings is the establishment of "seed banks." In Cuba, planting time does not always coincide with

Table 19
Three pesticide mixtures for stake treatment before storage or planting

Trade name	Common name	Rate (amount of commercial product per liter of water)
Formula No. 1		
Dithane M-22	maneb	2.22 g
Antracol	propineb	1.25 g
Vitigran 35%	copper oxychloride	2.00 g
Malathion W.P. 4%	malathion	5.00 g
Formula No. 2		
Malathion E.C. 57%	malathion E.C.	1.5 cc
Bavistin W.P. 50%	carbendazim	6.0 g
Orthocide W.P. 50%	captan	6.0 g
Formula No. 3		
Orthocide W.P. 50%	captan	6.0 g
Bavistin W.P. 50%	carbendazim	6.0 g
Aldrin 2.5%	aldrin	1.0 g/stake

the time of harvest, so the supply of fresh planting material is problematic. Rather than developing storage systems, the Cubans plant special seed production plots, or seed banks. These plots, which constitute about 10 percent of the total area planted on the state farms and cooperatives, are maintained in the field until the planting material is required, and only then are they harvested.

When seed banks are established, they can be given extra attention to ensure that good quality planting material is produced. For example, heavy doses of fertilizer may be applied to promote top growth, and chemicals may be used to control stem borers or other insects that reduce the quality of planting material.

Planting

Cassava can be planted throughout the year in areas with continual rainfall. Even in locations that have a 3- to 5-month dry season, cassava can be planted 2 to 3 months before the beginning of the dry season. The flexibility of planting time allows the farmer to spread his labor use. However, there is normally a considerable difference in yield associated with different planting dates, and the optimal planting date for high yields will vary depending on the local conditions. Thus farmers who plant at nonoptimal times to spread their labor will sacrifice yield somewhat.

The numerous ways of planting cassava have been exhaustively researched with somewhat conflicting results. Overall, vertical planting of cuttings 20 to 30 centimeters long produces yields that are equal to, or slightly better than, yields from other planting systems. A significant advantage of vertical planting is that the plants are less likely to lodge (topple over). Plants that are lodged not only yield less, they also sprout from the axillary buds of the stems, which makes them a poor source of planting material for subsequent crops.

Although most cassava is planted by hand, mechanical planters are available. In Brazil, a machine has been developed that takes long cassava stem pieces, cuts them, and then plants them horizontally. Its greatest disadvantage is its inability to plant vertically. Furthermore, the cutting blades may spread cassava bacterial blight to all the planting material. In Cuba, a prototype machine has been developed that receives long cuttings, cuts them, disinfects the cutting blades, treats the cut ends of the planting material with fungicides, and finally plants them in an inclined position. Many simpler tractor-drawn planters exist, and some can be adjusted to plant inclined. No mechanical planters have yet been developed that are suitable for use by small farmers who do not have tractors.

After planting, the major operations are weed control and, if required, control of pests through biological control agents. These topics have been discussed in previous sections.

Mixed cropping

Worldwide, at least a third of all cassava is planted in mixed culture. Cassava is commonly grown with other starch staples

Table 20
Yield of intercropped cassava compared with yield from monoculture in three countries

Cassava planted	Yield (t/ha) Cassava	Yield (t/ha) Grain legume	Intercropped cassava yield as percentage of monoculture cassava yield
Thailand			
Alone	27.6		100
With soybeans	26.7	0.69	97
With mung beans	26.4	0.77	96
With peanuts	24.5	0.91	89
India			
Alone	24.8		100
With peanuts	19.7	0.62	79
With cowpeas	16.6	2.03	67
Costa Rica			
Alone	16.8		100
With dry beans	15.2	1.45	90

Source: E. Weber, B. Nestel, and M. Campbell, eds., *Intercropping with cassava: Proceedings of an international workshop held at Trivandrum, India, 27 Nov.–1 Dec. 1978* (Ottawa, Canada: IDRC, 1979).

such as maize or plantains, but associations with rubber, coconut palm, oil palm, coffee, sesame, beans, cowpeas, or peanuts are not unusual. Mixed cropping of cassava with grain legumes holds great promise because of the potential nutritional benefits as the production of a high-protein grain will partially compensate for the extremely low levels of protein in diets based on cassava. In addition, the nitrogen-fixing ability of legumes may improve soil fertility. Planting cassava in combination with cowpeas or peanuts has been successful (see Table 20). Both of those grain legumes are relatively tolerant of light, acid soils, and their temperature requirements are comparable to those of cassava.

The best results with intercropping are obtained by changing the spatial arrangement of cassava plants in the field. Researchers in Brazil have developed a planting pattern in which two rows of cassava are planted close together with a wide space between pairs of rows, and at CIAT it has been found that cassava can be planted with a large degree of rectangularity—close together within rows but with widely separated rows (1.5 to 2 meters apart). Yields from both these systems equal those from more normal planting patterns, but in the early stages of cassava growth, there is sufficient space to grow a legume crop. If the two crops are planted simultaneously, the cassava competes with the grain legume for light by growing slightly taller and hence is not shaded out. By the time the cassava leaf canopy begins to close up, the rapidly growing grain legume has begun to ripen, and its leaves wither.

In this manner, high yields of both the cassava and the grain legume are achieved, and the advantages to the farmer are substantial. Mixed cropping suppresses weeds and lowers the disease incidence and pest populations in both crops. In addition, the grain legume harvest gives the farmer an early return on his investment, alleviating a major problem with cassava growing—the long time from planting to harvest, which frequently means that the farmer must obtain long-term credit.

Cassava is planted among mature coconut trees in India and among rubber trees in China, where crop land is extremely scarce. Under trees, the cassava tends to suffer from insufficient sunlight, and the harvest index is very reduced. The Central Tuber Crops Research Institute is attempting to develop cassava varieties that are adapted to this system. Yields are never likely to be high under trees, but the system results in some extra carbohydrate without requiring extra land.

Harvesting

Although cassava has no specific period in which it matures and ripens, there is an optimum period for harvest, which varies from variety to variety. If a variety is harvested early—before its optimum period—yields will be low; if harvested late, the starch and dry matter content may be low. Furthermore, at the onset of the rains following a dry period, starch content declines

A simple lever designed at IITA to make harvesting easier. (*Source:* R. Wijiwardene)

dramatically. Farmers often determine whether the starch content is high enough for harvesting by running a thumbnail across the cut surface of a root. When the starch content is high, a milky liquid oozes out. For farmers accustomed to this crude test, it is remarkably effective.

The crop is usually harvested by pulling the roots from the ground. When harvesting conditions are difficult, a crude lever is sometimes lashed to the stem. To reduce harvesting time and eliminate much of the backbreaking work, IITA has developed a simple harvesting aid that grasps the stem as it is raised.

Various tractor- and animal-drawn plows have been designed to turn the roots out of the ground or at least to loosen them to make hand pulling less arduous. A tractor-drawn device designed at CIAT loosens the roots and leaves them on or close to the soil surface. The roots can then easily be picked up by hand, separated from the stems, and packed for transport. This system causes less damage to the roots than traditional manual harvesting methods.

Completely mechanized harvesting systems have been developed in which the tops are pulverized and the roots lifted and deposited directly onto trailers for transport from the field. The best of these systems cause little root damage and could greatly reduce labor requirements for harvest in areas in which labor is expensive or scarce; they are not widely used, however.

RESULTS WITH NEW TECHNOLOGY

The new technology—the combined application of improved practices—has been tested under farm conditions in an area of the north coast of Colombia where soil and climatic conditions are adverse and farmers' normal cassava yields are 4 to 8 t/ha. The technology consists of

- careful selection of planting material
- treatment of the planting material with fungicides and insecticides (at a cost of less than US$10/ha)
- vertical planting at the optimum plant population (on fertile soils, with vigorous branching types—7000 to 10,000 plants per hectare; on less fertile soils or with less vigorous, more erect varieties; up to 15,000 plants per hectare)
- hand weeding, as needed, during the first 4 months after planting
- ridging in heavy soils in areas of high rainfall
- optional application of fertilizer

With no chemical pest control after planting, these practices raised yields of the local clone to 10 to 15 t/ha from 6 to 8 t/ha in the traditional system, and with a new hybrid, yields reached 20 to 30 t/ha. (Ridging is not necessary on the north coast of Colombia because the soils are sandy and well drained; however, in high-rainfall areas with heavier soils, planting on ridges is an essential component of the new technology.) The local clone showed no response to the application of fertilizer, probably because of its low yield potential and because in this area, a fallow system is used to maintain fertility. The new hybrid yielded more than the local clone in unfertilized fields and also responded to fertilizer. Tests in other areas have shown that when the fallow period is shortened, the local clones also respond markedly to the use of fertilizer.

QUARANTINE AND TRANSFER OF MATERIALS

The movement of diseases and pests is a danger not only between continents or countries but even within national borders. When cassava production is expanded into areas in which cassava has not previously been grown, careful attention should be given to quarantine measures. In a massive cassava project in the state of Minas Gerais in Brazil, planting material was brought from adjoining states, and yields declined rapidly as diseases and pests rapidly built up in the plantations. It is not certain that new problems were introduced with the planting material; however more care would undoubtedly have allowed the development of plantations that would have been relatively free of many of the diseases and pests that are now endemic in the area.

A technique developed by the National Research Council laboratory in Saskatoon, Canada, permits germ plasm to be exchanged internationally with considerable confidence that new pests or diseases will not be introduced as a result. The technique involves producing plantlets under sterile conditions from the apical meristem of cassava plants. This technique is now routinely used for the production of plants that are free of most diseases and pests. It has not been possible, however, to ascertain if African mosaic disease has been eliminated, even though heat pretreatment and culturing in this manner results in symptom-free plants. Hence it is not wise to import materials from areas that have mosaic disease (India and Africa). Moreover, there appears to be little justification for this exchange as African germ plasm has much less genetic variability than germ plasm from nearer the center of cassava origin in the Americas.

At present, many national quarantine authorities do not accept the transfer of vegetative material, even apical meristems cultured on agar in aseptic conditions. Ironically, quarantine restrictions on sexual seeds of cassava are much less stringent, though it is known that cassava bacterial blight is a seed-borne pathogen and hence may be introduced by importing seeds from infected plantations.

For moving planting material between areas or countries with similar pest and pathogen complexes, it may not be necessary to go to the trouble and cost of tissue culture. Vegetative material should, however, always be taken from well-managed plantations that are free of such systemic diseases as cassava bacterial blight, African mosaic disease, or frogskin disease. In addition, cuttings should be selected that have no obvious external or internal damage. The cuttings should be treated with fungicides and insecticides, both on dispatch and on receipt by the importing agency.

The knowledge gained over the last decade has made international germ plasm exchange a safe procedure if the norms and techniques described above are adhered to. Although there is always the danger that new pathogens or pests will be introduced, the safety inherent in the new methods means that the gains from germ plasm exchange far outweigh the minimal risks. Furthermore, if quarantine controls are excessively strict, there is always the danger of illegal importation, which entails no control on the part of the people who are determined to obtain the foreign material.

5
New developments in post-harvest technology

The extreme perishability of harvested cassava roots is a curse for everyone who grows, processes, markets, or consumes cassava. Some farmers who plant cassava solely for family consumption can avoid the problem by letting the cassava grow until it is needed, and then harvesting only as much as the household intends to eat immediately. In effect, the cassava is stored in the ground. But that is not possible for farmers who cultivate their land intensively and therefore must harvest their entire cassava crop at once to permit planting of the next crop.

Traditional means of storing fresh cassava usually involve burying it in soil or submerging it in water, but these methods make cassava virtually impossible to transport because it deteriorates rapidly when it is removed from storage. Leaving the stem attached to the roots, as some farmers in Ecuador and Colombia do, delays spoilage by a few days and makes transport somewhat easier.

When cassava is grown for sale as fresh roots, the middleman and retailer assume grave risks because the marketing process has to be completed within days. Supplies that are not sold quickly are likely to be lost. To cover the risk, a high marketing margin is necessary, and that makes fresh cassava costly for the consumer outside the production zone.

Processing is one way to reduce spoilage losses. Some cassava production areas have small mills that can rapidly produce dry flour or meals, but these mills operate inefficiently. The perishability of cassava keeps them from stockpiling enough roots to

For farm-family use, cassava can be stored in silos constructed from earth and lined with straw. (*Source:* CIAT)

run at a steady rate, and if production near the factory declines for any reason, supplies of roots from other areas are hard to obtain because of the complexity of harvesting, transporting, and delivering cassava before it spoils.

Deterioration also is a critical factor in processing cassava for animal feed. When cassava is sun-dried, the speed of drying depends on the wind and humidity. An unexpected shower can accelerate the spoilage of the roots waiting to be chipped and dried, lowering the quality of the final product. The perishability of cassava makes a continuous supply of cassava essential, which necessitates careful coordination between the farmers and the processors.

Although large industrial plants that process cassava into starch, alcohol, flour, or even *gari* can function irrespective of the weather, they, too, have difficulty operating at a steady rate. Their need for an orderly supply of roots compels the supplying farmer to harvest on a fixed schedule, which may not fit the availability of farm labor or the farmer's overall cropping plan. In addition, the supply may be disrupted when heavy rains make harvesting and transport from the field impossible.

Storage of fresh cassava

Post-harvest deterioration of cassava is related to two separate processes: physiological changes and microbial changes. Physiological deterioration often begins within 24 hours after harvest. The symptoms are blue or brown vascular streaking in areas of the roots just below the peel, or rind, which first becomes visible at the cut ends of roots or in damaged areas. Some varieties are slow to deteriorate, but they usually have a very low dry matter content, which makes them undesirable for either the fresh market or processing.

Microbial deterioration normally occurs later than the onset of physiological deterioration, but it often starts within a week after harvest. The signs are blue or brown streaks throughout the fleshy part of the root and soft spots. Spoilage is fastest in roots that are badly damaged during harvest or that are thoroughly soaked by rain.

Storage for human consumption

Cassava usually has to be consumed within a day or two after harvest because of the lack of satisfactory long-term storage methods. In some early work on this problem, the Tropical Products Institute (U.K.), in cooperation with CIAT, tested a system based on the European potato clamp. High humidity in the clamp "cures" the roots, which can completely eliminate physiological deterioration though microbial deterioration can still be a problem. A major disadvantage of this system is that the results are extremely variable, with complete loss occurring occasionally. That problem can be partially overcome by packing

Bulkiness makes it impractical to transport cassava roots over long distances. (*Source:* CIAT)

the roots in moist sawdust in boxes, which creates a high humidity environment like that of the clamp. The roots can be stored for up to 2 months, and the boxes can be transported, though they are bulky and heavy. But the method is costly for the farmer in terms of both labor and material inputs. Furthermore, it requires careful selection of undamaged roots, and it does not avert microbial deterioration.

In Trinidad and Tobago, it was found that roots packed in polyethylene bags could be stored for 2 months. Tests of this method in Colombia have shown that although physiological deterioration is prevented, microbial deterioration frequently causes complete loss after 1 or 2 weeks of storage. Dipping the roots in a fungicide before they are packed in the polyethylene

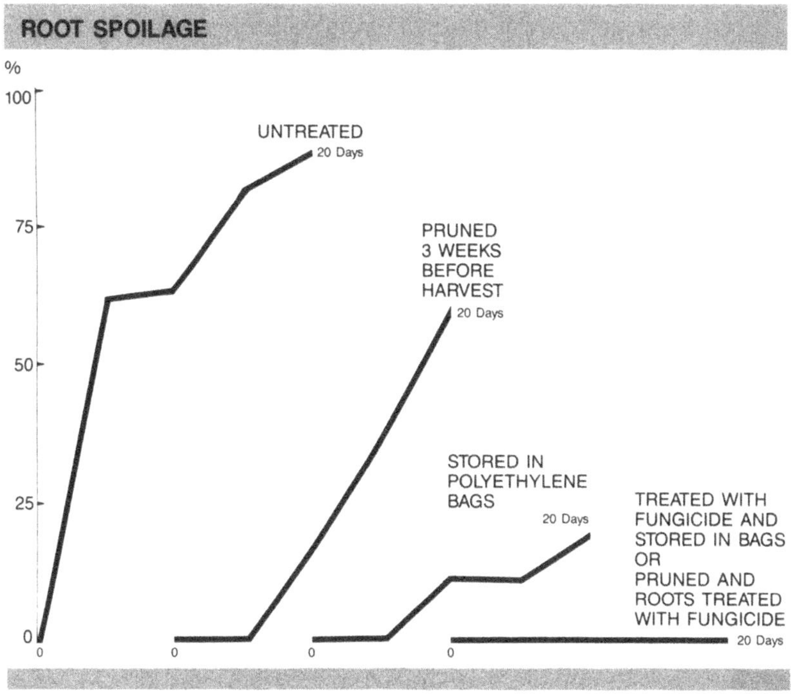

Figure 11. Spoilage is prevented for 20 days by applying fungicide to roots packed in polyethylene bags or to roots from plants pruned 3 weeks before harvest.

bags, however, can retard spoilage for up to 3 weeks (see Figure 11). Research to determine if the fungicide leaves any toxic residue must be completed before this technique can be recommended. Since the fungicide is not translocated and is applied to the surface, and since roots are always peeled before cooking, residues are not likely to be a problem. The spoilage retardant, Mertect (thiobendazole), has extremely low human toxicity, and, in fact, the product has been used as an antihelminthic drug for humans. Consequently, packing cassava roots in polyethylene bags after application of a chemical protectant may overcome many of the problems associated with the perishability of fresh cassava for human consumption. Although treating and packing the roots would be laborious, the reduction in the existing high marketing margins could make the practice economically sound.

The spoilage of fresh cassava can also be prevented by freezing the roots, which is done on a limited scale in the Dominican Republic and Costa Rica for export and in Colombia for transport to large urban centers. This method is extremely expensive, and in most developing countries, frozen storage of a basic starch staple is not economical.

Storage for processing

The physiological deterioration of cassava is related to an accumulation of the compound scopoletin. When the roots are bruised, the compound accumulates, and the roots become discolored. But pruning the plants (leaving only a woody stump to facilitate harvest) 3 weeks before harvest prevents physiological deterioration after harvest even if the roots are severely damaged. Pruning, however, causes changes in the root that lower its eating quality and lengthen the cooking time. Nevertheless, such roots are satisfactory for animal feed. The dry matter content, in fact, is increased, which is a definite advantage. It is not known whether roots from pruned plants are acceptable for starch extraction.

Pruning permits cassava roots to be held for approximately 5 days before microbial deterioration begins. The use of a fungicide to retard microbial deterioration in combination with pruning 3 weeks before harvest stretches the safe storage period to at least 3 weeks. This technique could greatly alleviate the logistical problems of the processing facilities by making it possible for them to maintain a steady supply of roots.

Chips and pellets

In Brazil, oil- or wood-fired driers are used to produce cassava chips, or "raspas," that are of such good quality that they are milled into flour for making bread and other bakery products. The high price of energy has, however, caused most processors to shift to sun drying. Sun-dried chips are often of poor quality because the roots, which start to deteriorate before processing even begins, continue to spoil during the 3- or 4-day drying period and because of contamination. The chips may also further deteriorate in storage if inadequate drying leaves the moisture content too high. However, recent advances in chipping and drying technology have improved both the efficiency of drying and the quality of the final product.

Although much less bulky and easier to transport than fresh roots, cassava chips are a bulky commodity (around 400 g/1). In order to lower chipping and handling costs, Thai exporters have moved toward pelleting cassava chips, which increases their density (to 680 g/1). An added advantage of pellets is that during handling, they create less dust than chips.

Chipping

The shape of the chip influences how fast it dries. Because cassava chips are white, they reflect much of the sunlight that falls on them. Hence in sun-drying systems, the chips are dried more by the passage of air over them than by the direct effects of the sun's rays. The initial drying of the chips occurs as the water on the chip surface evaporates. As the surface becomes drier, water vapor diffuses from the inner parts of the chips. Although diffusion and the rate of drying are fastest in small chips, they easily become compacted, which prevents free air movement. As a result, for effective drying, the chip's shape should permit air to readily pass through a large mass of chips. The optimal chip geometry for natural drying is a bar approximately 5 x 1 x 1 centimeters.

In Malaysia, a simple machine has been designed that produces chips of approximately those dimensions. This chipper has been modified by several organizations and can be built by workshops with locally available materials. In Mexico, a chipper with a throughput of 1200 kilograms of fresh roots per hour costs approximately US$700. These efficient chippers, which are now widely used throughout Malaysia, produce much higher quality chips than those produced by machines in Thailand, although the Thai-type chippers have now been modified to yield higher quality chips. A drawback of the Malaysian type of chipper is that manufacturing the cutter blades requires highly skilled blacksmiths.

Malaysian-designed chippers produce cassava chips with the optimal shape for fast drying. (*Source:* CIAT)

Drying

The common method of drying chips is to spread them in a thin layer on a large concrete floor. The drying efficiency of this method is low because most of the sunlight is reflected by the chips. Painting the concrete surface black does little to increase the absorption of solar energy, as the concrete soon becomes covered with white cassava starch and flour, and when the chips are spread out, they reflect the sunlight before it can be absorbed by the black surface.

Research by the Asian Institute of Technology, the Tropical Products Institute, and CIAT has shown that drying rates can be greatly enhanced by placing the cassava chips in suspended

Placing chips in raised trays accelerates drying. (*Source:* CIAT)

trays. The trays allow more air to flow through the mass of chips and eliminate the need to turn the chips periodically. The capital cost of the trays is relatively high per unit of dried cassava because each tray usually can only dry one batch of cassava every 2 days.

With careful scheduling, the time needed to dry cassava chips can be held to 24 hours. Cassava chips made during the day should be loaded onto trays in the late afternoon. During the night, even though the relative humidity may be high, the chips lose moisture. On the following day, the difficult final drying to 12 to 14 percent moisture content occurs during the early afternoon when the relative humidity is lowest. This schedule reduces the capital cost per ton of cassava dried in trays to a level that may make the method competitive with drying on concrete floors, and the resulting higher quality product can be expected to command a premium price. Large-scale trials have not yet been made to test this technology on a commercial basis, however.

Small-scale flour and starch production

Cassava flour, such as *gari* or *farinha,* is usually produced in small processing plants or by the farmer using simple equipment at home. In traditional processing, cassava is manually peeled and sliced or rasped before drying. Since all these processes are arduous, a variety of small machines that can be built by blacksmiths have been developed to wash, peel, and rasp fresh cassava. These machines have been widely adopted in Brazil for the production of *farinha,* and they are becoming popular in West Africa.

Large starch plants with centrifugal separators and flash driers produce an extremely high quality product and extract more than 90 percent of the total starch in the cassava roots. There are several international construction companies that will design and install complete starch factories. Small-scale starch factories normally produce a lower quality starch with more impurities and have much lower extraction rates. But washing the roots after peeling and before rasping can greatly reduce the level of impurities in the resulting starch.

Alcohol

Using cassava as the raw material to produce fuel alcohol is not new. Before and during World War II, Australia grew cassava as raw material for alcohol production, and during the war, Brazil produced more than 60 million liters of fuel alcohol per year from cassava. The low cost of petroleum products in the 1950s and 1960s ended these activities, but with rising oil prices, interest in cassava alcohol has been renewed.

The basic process is to wash the cassava and mash it. The mash is heated to gelatinize the starch, which is then liquified by addition of the enzyme alpha-amylase. The liquified starch is hydrolyzed to sugars, and the sugars are converted to alcohol by yeast fermentation. The resulting liquor has a low alcohol content, which has to be raised by distillation. Present technology allows for an extremely efficient recovery of alcohol: The commercial yield of the first modern commercial cassava alcohol plant at Curvelo in Brazil is 170 liters of alcohol per ton of fresh cassava.

The conversion of starch to sugars, saccharification, can be done by acid hydrolysis or biologically through the use of amylolytic enzymes. The acid processes degrade the sugars and result in low levels of alcohol production. In Brazil, the acid processes have been replaced with microbially produced enzymes, which are more effective.

A major problem with producing alcohol from cassava is the energy used in the processing. When alcohol is produced from sugarcane, the waste stalks (bagasse) are burned to fuel the distillation process. Although cassava stems could be burned, they would not provide sufficient energy for distillation unless the fermentation process were further developed to raise the alcohol content in the fermentation liquor. Considerable attention is being given to the development of more efficient processes to separate the alcohol from the water.

6
Some case histories of cassava development

During the last several hundred years, there have been numerous attempts to promote cassava production. Some have failed, some succeeded. In Asia, for example, cassava growing has resulted from the purposeful introduction of the crop, and the large areas now planted to cassava demonstrate the success of these introductions. Because of the profound political and economic changes developing countries have undergone in the past 20 to 30 years, it is not worthwhile to attempt to draw lessons from programs of the colonial period. Instead, this chapter gives five examples of recent approaches to the development of cassava production and utilization under very different circumstances.

Kerala, India: Improved production for local consumption

With approximately 25 million people, the state of Kerala in southern India is more populous than many countries. Population density is extremely high, farms are tiny, and good agricultural land for production of rice, the major staple, is scarce. Most of the cultivated area is either poorly drained lowlands, where the important crops are rice and coconuts, or well-drained hilly areas, where cassava is grown. The soils of the hilly areas are highly weathered, very acid, and low in fertility.

Expanded food production is needed to improve the nutrition of the population. Rice, cassava, and coconut predominate in

Table 21
Average daily per capita food energy and protein supplied by staples in Kerala, India, 1970/71 and 1976

	Food energy (cal)		Protein (g)	
Staple	1970/71	1976	1970/71	1976
Rice	1000	689	18.5	13.7
Cassava	744	827	3.3	5.5
Coconut	267	388	2.7	
Oil	212	128		
Sugar	100			
Other	196		13.3	22.3
Total	2519	2032	37.8	41.5

Sources: 1970/71: United Nations, *Poverty, unemployment, and development policy: A case study of selected issues with reference to Kerala* (New York, 1975); 1976: S. K. Kumar, *Impact of subsidized rice on food consumption in Kerala*, Research Report no. 5 (Washington, D.C.: International Food Policy Research Institute, 1979).

the diets (see Table 21), which are deficient in both protein and calories. During the 1950s and 1960s, the proportion of cassava in the diet increased significantly, with no change in the total calorie or protein intake and a decline in the importance of rice. This change suggests that as the population has expanded, cassava production has grown at an even greater rate in response to rising food needs while rice production has not been able to keep pace.

Cassava production in India (most cassava is produced in the state of Kerala) has expanded dramatically since 1950, largely because yields have risen from 5 t/ha to 16 t/ha. These yields are among the highest in the world even though Kerala has infertile soils and a shorter growing season—less than 12 months—than many other cassava growing areas.

Research work on cassava in Kerala has a long history. The Central Tuber Crops Research Institute (CTCRI) conducts research that is oriented toward the development of improved agronomic practices and the production of improved varieties. As the institute is in the heart of the cassava growing area, the technology developed is soon adopted by nearby farmers and, due to the high population density and the proximity of farms, the new technology spreads rapidly. A contributing factor in the rapid technology transfer has been the interest of local religious leaders in the research work and its application in farmers' fields.

The introduction of several varieties from Malaysia in the 1950s was a technological breakthrough for Kerala. The varieties were extensively tested, and among them, M-4 was found to have moderate yield but extremely good cooking quality. M-4 rapidly became the predominant variety planted by farmers. Although higher yielding hybrid clones have since been developed, none have had sufficiently good cooking quality to replace M-4. Several new high yielding, good quality clones are now being tested in farmers' fields.

At the same time that M-4 was being introduced, CTCRI was developing improved agronomic practices. Recommendations, such as the size of cuttings, time of planting, form of planting, and weed control, that involve few purchased inputs appear to have been widely adopted by farmers. Other recommendations, such as application of lime and fertilizer, have been less widely adopted. However, most farmers apply farmyard manure, which is readily available and which research has shown to have great benefit for cassava.

African mosaic disease is a major problem in Kerala, and CTCRI has an ambitious "seed" program to deal with it. Some years ago, CTCRI began to eradicate the disease on its experimental farm so that it could distribute stocks of uninfected planting material. CTCRI found that when clean planting material is used, reinfection rates are extremely low, even on plots surrounded by infected fields. Clean planting material is being distributed to farmers, and the results are promising. It appears that with the new hybrids and clean seed, there are great possibilities for continuing yield increases.

Since diets in Kerala are short of calories and the population continues to grow, the demand for calorie sources normally exceeds the supply. Thus, for example, in years when rice is in short supply, the demand for cassava tends to increase. In addition, the low income level makes the population sensitive to price differences in basic food staples. Per calorie, cassava is normally about half as costly as rice bought on the open market (about one-fifth of the total calorie intake in Kerala comes from subsidized ration rice, which, per calorie, is similar in price to cassava—however, the quantities of ration rice are limited), and this lower price tends to result in a strong demand for cassava. Thus the new technology has managed to enlarge production, and prices have remained low enough for local demand to be able to absorb the increased supply. Farmers have benefited from increased production while consumers have gained a larger supply of cheap calories.

Caicedonia, Colombia: A regional program to provide food for a major urban center

Caicedonia is a rural community with 126,000 inhabitants and 24,000 hectares of farmland situated in the mountains of Colombia. The conditions are close to ideal for cassava: The soils are fertile, but somewhat acid; the total rainfall, about 1900 millimeters, is well distributed throughout the year; and the average monthly temperature varies little from 22.5° C. Historically, however, the area's economy was based on coffee and plantains, and until the late 1960s, only 60 hectares were planted to cassava.

The powerful and well-organized coffee growers' federation (Federación Nacional de Cafeteros) is the driving force behind the development of the area. This group has assisted development in the zone through credit and marketing schemes, electrification, construction of roads, and education programs. The overdependence of the area on coffee, which places the economy at the mercy of volatile world coffee prices, has been one of the main concerns of the federation. As a result, the organization supports a large diversification program aimed at promoting and developing crops other than coffee.

The coffee growers' federation began to investigate cassava growing in Caicedonia and the surrounding areas in the late 1960s. Marketing studies suggested that there was a large potential market for fresh cassava in the capital, Bogota, and that a few farmers were obtaining yields as high as 15 t/ha. Observations of local varieties showed that one, Chiroza Gallinaza, yielded well, but more important, its roots had good eating quality and a long post-harvest shelf life.

In 1971, the federation selected a young agronomist to spend a year at the Centro Internacional de Agricultura Tropical (CIAT) undergoing intensive training in cassava production methods. During this period, he was actively involved in varietal selection, development of agronomic practices, and pest control measures. After completing his training, he returned to his post in the Caicedonia area, where the federation provided a jeep and financial support so that he could visit farmers. He advised on production technology and helped farmers obtain credit from the federation for cassava production. The federation gave credit only to growers who adopted the recommended technology.

A 1973 survey of cassava production in the zone revealed the impact of this extension effort: The yields of farmers who received credit and used the recommended technology were 25 t/ha; those who did not, had yields of only 8 t/ha. The area planted to cassava also increased rapidly—500 hectares were in cassava by 1974 and 1300 hectares by 1978. This success prompted farmers in neighboring municipalities to plant more cassava, adding 4000 hectares to the region's cassava area.

Although sudden production increases in other areas of Colombia have often led to severe marketing problems, forcing reductions in the cultivated area, this situation did not occur in Caicedonia as the marketing system that evolved insulates the farmer from vulnerability to day-to-day variations in the market supply of cassava. A small number of middlemen buy the cassava as a standing crop. These middlemen estimate the production of a field and offer the farmer a price for the whole field. If the farmer accepts, the middleman takes responsibility for clearing the field by a certain date. Each middleman has his own harvesting team and trucks and schedules the harvesting and transport to provide a steady supply of freshly harvested cassava to the Bogota market. As a result, cassava from Caicedonia has a reputation for always being fresh, and it now commands a substantial price in Bogota. Although the marketing margin is high and the middlemen's profits are large, it appears that this system works to the benefit of both the farmer and the consumer. The farmer does not have to use his valuable time studying the daily market and peddling his crop, and the consumer obtains a high quality product.

Most problems of the Caicedonia project have been of a technical nature, and the proximity of the national research agency, the Instituto Colombiano Agropecuario, and CIAT has permitted a quick resolution of many difficulties. For example, cassava cropping led to an outbreak of root rots soon after the project was begun, an outbreak that was aggravated by poor drainage in some fields. Since the farmers were reluctant to rotate with a less profitable crop, the scientists suggested planting on ridges as a way to reduce the incidence of infection. Several farmers adopted the practice, and as the results became known, the practice soon became universal in the region.

Shortage of planting material was another problem. In 1974, when the cassava land was expanding rapidly, cassava growers in the area were unwilling to sell planting material to new producers as they feared that overproduction might result and reduce prices. Consequently, one farmer who had been refused went to another area to buy planting material. The cuttings he purchased were infested with bacterial blight, a disease not found in Caicedonia, and his plantation soon became heavily infected and threatened the surrounding fields. Scientists recommended destroying the diseased field. The owner agreed to harvest his field and destroy all the residue if the other farmers would sell him planting material. In this manner, bacterial blight was eliminated, and the local farmers became more amenable to trading planting material.

Despite these successes, yields have declined in recent years. The long-term consequences of poor management on disease and pest incidence have been stressed at field days, but it is difficult to convince growers who are getting very high yields to change their practices. Some farmers, in their eagerness to increase their plantings of cassava, used cuttings infested with the cassava fruit fly (*Anastrepha* spp.) and frogskin disease. Moreover, few farmers rotated their crop before yields became very low, and most farmers applied massive levels of insecticide at the first sign of an insect attack, killing not only the pests but also their natural enemies. All of these factors led to serious pest outbreaks in the area. Farmers now are more aware of the long-term consequences of poor pest management and are beginning to employ biological control, to select planting material carefully, and to rotate crops.

These problems have not kept the cassava industry from expanding rapidly around Caicedonia. From an insignificant part of the economy in 1970, it has grown into an enterprise that earns US$8 million per year and is a major factor in the rising prosperity of the region.

Cuba: Large-scale production for food

Since 1960, Cuba's policy for meeting its food needs has been to employ highly mechanized, high-input technology on large state farms. Although agricultural output has increased, the rapid rise in the cost of chemical inputs has prompted a search for alternatives. In the 1970s, the Centro Experimental de Mejoramiento de Semillas Agamicas (CEMSA) began work on cassava, a traditional food in Cuban diets. Yields were exceptionally low (approximately 4 t/ha), numerous cassava varieties were grown, and little was known about their relative value for agricultural production. As a first step, CEMSA collected and evaluated clones from Cuba and southeastern Mexico. Two varieties, Senorita and Pinera, were selected to be multiplied for large-scale trials on state farms. Senorita was the highest yielding but was very susceptible to bacterial blight; Pinera had a slightly lower yield potential but was relatively tolerant of bacterial blight.

A critical question was how to multiply these clones rapidly and distribute them without spreading bacterial blight. Scientific reports suggested that the incidence of bacterial blight might be less severe in the lower or primary branches, so these branches were selected for multiplication, and as a result, the incidence of the disease has been greatly reduced.

CEMSA also began to test new growing practices. Cassava management in Cuba in the 1970s was based on sugarcane production practices. Thus cassava was planted horizontally in deep furrows and irrigated frequently. Experiments quickly showed that planting on the top of the ridge in a vertical or an inclined position and less-frequent irrigation would greatly increase yields. A related problem was mechanization, since Cuba, unlike most developing countries, has a shortage of agricultural labor. The development of a machine that places planting material in an

Carefully produced cuttings being distributed at a state farm in Cuba. (*Source:* CIAT)

inclined position on the ridges paved the way for large-scale production of cassava.

As a result of further development and consultation with CIAT, the Colombian system (*el sistema colombiano*) of cultivation was recommended. Most state farms and cooperatives now use this system. It consists of planting special "seed banks" using certified planting stocks, a careful selection of planting material from the primary branch, treatment of the planting material, planting on the top of ridges, less-frequent irrigation, and reduced insecticide application. Yields average 18 to 20 t/ha with some state farms obtaining up to 30 t/ha. Cuban officials claim that total production has doubled as a result of the use of the Colombian system and the clones Senorita and Pinera. An early maturing Brazilian clone, Mantiqueira, introduced from Colombia, has been released and is grown widely, extending the period of the year when cassava is available. The increased output is being marketed through state channels.

Thailand: An export crop

Thailand is one of the few developing countries that is a net exporter of cereal grains. Rice has been both the country's major food crop and its main export. In 1950, rice accounted for 70 percent of the country's agricultural output; since then, considerable emphasis has been placed on diversification. Crops such as sugarcane, maize, and cassava have increased in importance, and rice now accounts for less than 50 percent of production.

Marketing

Cassava production in Thailand expanded from 3 million tons to nearly 18 million tons during the 1970s, largely as a result of trade agreements governing the importation of cereals to the European Community. Since 1967, the European Community has protected its cereal growers by a levy on imported grains. The levy is set at a level that is commensurate with farm income objectives, and it has lifted cereal grain prices well above world market prices. However cassava, which is imported for feed, is subject to a levy of only 18 percent of the levy on imported barley or 6 percent of the value of the pellets or chips, whichever is lower. This policy has kept the price of imported cassava chips and pellets low in relation to that of cereals and has been of fundamental importance in the development of the Thai cassava trade.

In 1950, cassava occupied about 0.2 percent of Thailand's total crop area, and cassava exports were almost exclusively to the United States in the form of starch. In 1956, European shipping and trading companies introduced the by-products of cassava starch manufacture to European animal feed markets. A few years later, the development of small-scale chippers driven by gasoline engines made it practical to produce dried cassava chips directly from fresh roots. Thousands of small chipping plants with concrete drying floors were constructed, and chips completely replaced the by-products in the export market.

Although cassava chips are a cheap source of calories, they are expensive to transport. They are bulky and fragile, and they produce clouds of dust during handling. In the late 1960s, German entrepreneurs invested US$1 million in the first Thai pelleting plant and began shipping cassava pellets to Europe. Compared with chips, pellets are much less bulky (and almost equal in density to cereals) and less dusty, though dust still causes difficulty in handling. Thailand's exports of pellets now greatly exceed its exports of starch. By 1980, more than 400 pelleting plants had been built, and exports had reached 6 million tons worth $550 million.

The growth of the cassava industry was spurred by the development of an extensive road network for strategic reasons. The pellets are hauled by truck to the major ports where they used to be loaded onto small cargo vessels for shipment to Europe. In the 1970s, rising fuel costs made the use of 100,000-ton-capacity ships economically attractive for transporting cassava, but Thailand did not possess deepwater port facilities to handle these large ships, so they anchored in the Gulf of Siam, and bagged cassava pellets were delivered on barges. This unwieldy system ended when the world's largest conveyer-belt loading pier was built at Mabookrang in 1977, which can load 100,000 tons directy from shore in less than a week.

The Thai trade has, however, run into certain problems. European importers stipulate a minimum quality requirement of 62 percent starch, 5 percent crude fiber (maximum), 3 percent sand (maximum), and a moisture content of less than 14.0 to 14.3 percent depending on the season. Due to the crude drying technology used, there is considerable fungal and bacterial contamination of the chips, and the starch content is often low. Furthermore, instances of adulteration with sand have sometimes further lowered the quality. In a survey of material entering Holland from Thailand in 1971, more than 80 percent of the samples failed to meet the minimum requirements. Belgium has stricter quality requirements than the other importing countries, and it has increased its imports of higher quality chips and pellets from Indonesia at the expense of those from Thailand. Thai officials, concerned about the quality of their country's export products, have put the burden of meeting requirements on the exporting companies instead of relying on government control, as was the case previously. To demonstrate the gravity of not obeying the regulations, a Thai merchant was recently sentenced to life imprisonment for adulterating cassava with sand.

Aside from quality problems, the Thai cassava industry faces uncertainty about the continuation of tariff agreements that make the pellets attractively priced in the European market. The European Community could raise the levy on Thai cassava pellets at any time, and there is considerable pressure from European farmers to do so. Thailand has accepted export quotas that will gradually decrease cassava exports to the European market over a period of years, which clouds the future of the Thai export trade in cassava.

The European Community has offered US$40 million to help Thailand grow crops other than cassava. Before cassava production spread to northeastern Thailand, the major crop was kenaf, but it never occupied more than a fraction of the land that is now in cassava. The expansion of cassava has brought limited wealth and development as well as political stability to the area, and at present, there is no realistic substitute for it. The system of quotas currently in effect is driving down the price the small farmers receive and reducing their standard of living.

Production

Before the development of the pellet industry, cassava was mainly produced in southeastern Thailand. As the demand soared, there was no room for expansion in that traditional area, so cassava spread into the northeastern and southern provinces. The good market, plus the excellent local variety Rayong 1, made cassava an attractive crop for small farmers, particularly in areas with poor soil and no irrigation facilities.

Banks and merchants supported the expansion of cassava plantings by making credit available. Some merchants also began to ship large quantities of planting material from the traditional growing areas to the small farmers in the new areas, and the merchants advised the farmers about suitable growing methods for cassava. Also, as cassava production grew, village entrepreneurs constructed a large number of chipping and drying plants, which used cheap, locally made machinery and had a heavy labor requirement. These plants were able to buy all the cassava produced. Thus, with credit from the banks and merchants, with the availability of inputs, and with an assured market, little else was required to stimulate the small farmer to increase cassava production.

Although the average cassava yields are high, 15 t/ha, there are some signs of decline. The Thai cassava industry is based on a supply of cheap fresh roots, which in turn is dependent on good yields, so more farmers must begin to use fertilizer and improved technology in the future or the Thai industry could lose its competitive advantage.

The north coast of Colombia: Chips for the local feed market

The north coast is a major cassava producing area of Colombia, with cassava occupying 40 to 50 percent of the cultivated land on farms of less than 20 hectares. Until the late 1970s, almost all the cassava produced in the area was consumed on the farm or sold fresh in local or urban markets. A small starch plant in the area absorbed less than 5 percent of the production.

In 1976, Colombia established an integrated rural development program (Desarrollo Rural Integrado, or DRI) to increase the production of food staples, improve the welfare of the small farmers, and assist them in purchasing inputs and marketing. The DRI program was to coordinate the activities of government agencies responsible for agricultural credit, farmers' organizations, national training, agricultural research, and credit for small-scale processing plants. On the north coast, the departments of Sucre and Cordoba were selected as target areas for the first phase of the project.

Initial attempts to increase cassava production were not an unmitigated success. Although production increased as credit lines were opened to the small farmers, the fresh market soon became saturated, prices fell drastically, and many farmers did not even consider it worthwhile to harvest their cassava. As a direct result, DRI set up a post-harvest committee to deal with the marketing of the various crops it was working on.

At about the same time, an independent study was made on the potential of cassava chips as a substitute for imported cereal grains in the local feed industry. At the level of prices paid for fresh cassava for human consumption, a cassava-chipping industry did not appear viable; however the analysis suggested that if yields over 8 t/ha could be obtained with the existing production costs, then cassava could profitably be produced for animal feed.

But this analysis was based on many assumptions and certainly could not, in itself, be considered a sound basis for setting up a scheme for increased production and drying throughout the region. Two alternative lines of development were considered. The first was to embark on a lengthy research program to verify the assumptions on which the analysis was based; the second was to set up a small pilot project, using the existing knowledge about cassava production, to see if cassava chips could be economically produced. The second alternative was chosen.

In 1980, a group of small farmers in the department of Sucre constructed a drying floor of 300 square meters, and in the following year, the pilot project began to collect data on drying under the local conditions and to test the acceptability of the product to the local animal feed industry. As experience was gained, researchers modified and improved the drying process. The actual drying was managed by the farmers' group, so accurate information on the labor requirements and costs of the process could be obtained. DRI induced the largest local feed mill to guarantee a price for the dried cassava chips. In 1982, the chipping plant operated on a semicommercial basis, and further data were collected. In 1983, encouraged by the pilot project, DRI decided to help other farmers organize formal farmers' associations, construct drying floors, and purchase chipping machines and the small equipment used in drying cassava. Six chipping plants began operating in 1983, and 20 more are planned.

When the first six commercial plants were under construction, the project organizers were concerned whether enough cassava would be available to keep the plants fully operational and whether a possible shortage of roots in the area would drive up cassava prices (for the local fresh market), keeping the drying plants from obtaining low-priced raw material. The farmers, who also owned and managed the chipping plants, were not, however, concerned about a high cassava price. They felt that if prices rose, they could sell on the fresh market at a high price and cover the interest on their debts. On the other hand, if the fresh-market price fell, they could process their cassava and sell at the price guaranteed by the local feed mill. Hence, for the first time, the farmers could think of increasing production with no risk of flooding the market and having to sell their cassava at ridiculously low prices. Consequently, they responded by planting more land to cassava and improving their production technology.

Although only a few plants are functioning at present and their total production is not large, this project, which integrates production, processing, and marketing, does seem to offer several hints on how successful cassava expansion schemes can be managed.

Successful development: An appraisal

These five disparate examples demonstrate that there is no simple formula for success. In three of the cases, development of cassava production appears to be closely related to the improvement of technology by research organizations, but in one case, that of Thailand, though much has been done in the past, research is only now beginning to play an important role. Nevertheless, the existence of a technology that produces high yields appears to be of paramount importance. In four of the five cases, yields are well above the world average.

A significant common feature is the presence of a stable market for the cassava produced. Although the marketing systems vary—a complex chain of processing and shipping in Thailand; a local feed market in northern Colombia; a state distribution system in Cuba; a middleman marketer in Caicedonia, Colombia; and small local markets in India—in all cases the producers have an outlet for their produce that does not require them to devote much time to marketing. In addition, the farmers have been able to rapidly obtain information about new production technology. In the examples, the transmission of this information was through channels as different as extension agents, merchants who gave advice, and a state-controlled production system. However, in all cases a channel existed.

Although many other factors will affect the adoption of improved cassava technology, it does appear that successful programs for increased production must have at least three components: a viable production technology, a way to communicate this technology to the farmers, and an outlet for the increased production. The experience on the north coast of Colombia and observations of several cassava projects that have failed suggest that certain ingredients are essential for a successful implementation of cassava development programs.

The first phase in a development project is a macroeconomic analysis of the demand for cassava or cassava products and competing products, such as cereal grains, for which cassava can substitute. In this analysis, not only the potential demand for cassava products but also the target prices at which cassava could enter the various markets should be determined.

When potential demand has been established for one or more cassava products, production sites and processing technology should be evaluated in terms of both potential output and production and processing costs. These estimates may be relatively crude, but they can be compared with the target prices to decide whether to continue with the cassava project.

If the decision is positive, two lines of approach can be followed. The original crude estimates can be refined by further research, but this task may be costly, and the extra accuracy gained may be small. Or a small pilot project can be started using the best available production and processing technology to obtain concrete data on a semicommercial basis. During this phase, technology will be adapted and improved by research at the site. This approach will test the technological and economic viability of expanding the project into a commercial phase.

Certain policy decisions may also be required. For example, it may be necessary to establish special credit lines, increase institutional support of extension work, or remove subsidies to competing products. Once the policy changes have been instituted, the project can be moved into the commercial phase on a sound basis with great expectations of success.

7
Development of national programs

Most tropical countries are short of calories for human consumption or animal feed, and despite the tremendous potential many of these nations have for growing cassava, few have well-organized efforts to expand cassava production and utilization. Cassava producers are often hesitant to increase their production because they fear that they will flood the market. Food processors, the animal feed industry, and other possible industrial users are unwilling to invest heavily in cassava-based products because the supply of cassava is often erratic or insufficient for their requirements.

Both growers and processors often have good reasons for not changing the status quo. As described in Chapter 6, attempts to increase cassava production on the north coast of Colombia were not initially successful. The perishability of the roots kept the farmers from transporting their surpluses to distant markets where cassava prices were favorable, and farmgate prices declined so sharply that the farmers could not cover the costs of planting and harvesting the crop. At the same time, prices for maize and sorghum for animal feed were extremely high. Cassava chips could have competed well in the local market for feed, but chipping and drying plants did not exist and could not be constructed in time to utilize the excess production. The reverse situation occurred when a large drying plant was constructed in Venezuela to process cassava for animal feed. The plant's capacity was so much greater than the local production that it could not work at more than half its full capacity.

These two examples underscore the necessity for carefully planning the integration of production, processing, and marketing in the development of a cassava program. Many agricultural leaders regard production as the primary bottleneck in agricultural development, and for most crops, that assumption is basically true. For instance, in Central America, most nations have campaigns to increase the production of dried beans. These programs are little concerned about overproduction because crops of such relatively low perishability can easily be transported to other areas or even exported. With cassava, a highly perishable crop, excess local production is a serious problem. Any program to expand output must be integrated—it must address the processing and marketing aspects of the crop as well as production.

In establishing integrated cassava programs, a major question is which agencies will be involved in the research and development. In developed countries, a large portion of the total agricultural research and development budget is borne by the private sector. Private companies may develop varieties and distribute seed, produce chemicals that help to increase production, and devise new methods of packaging and processing to enhance the salability of the final product.

Some similar examples in regard to export crops exist in the developing world. In Malaysia, a consortium of companies that produce palm oil has invested heavily in research. The total production of these companies is so great that they can obtain a high return on their investment in research because of the resulting increases in productivity. But, with the exception of Thailand, cassava is not a major export crop, and even in Thailand, most of the cassava is produced by small farmers. Hence it seems unlikely that large private companies will engage in cassava production and processing and give support to research and development.

In the developing countries, particularly in Latin America, the private firms have invested in research and development when there is a possibility that the product being developed can be marketed. For example, private companies have bred sorghum and maize hybrids that are adapted to lowland tropical conditions; since farmers cannot produce their own hybrid seed, the companies can sell the seed of a good variety. In regard to cassava, most farmers produce their own planting material, and they can purchase or obtain small stocks of new clones and multiply

Farmers leaving a field day carrying cuttings of new varieties to try on their own farms. (*Source:* CIAT)

them themselves. Hence the "seed" market is too small for the private sector to be interested in developing new clones.

The development of chemicals specifically for use in cassava production also appears unlikely, because the low-input philosophy, which seems appropriate for cassava, negates the possibility of large sales. However, it does seem probable that cassava farmers will have to use fertilizers and it will be convenient for them to use herbicides in the future, so commercial companies may become interested in investigating these aspects of cassava production. Most cassava research, however, is not going to be economically attractive to the private sector. It will have to be done by the public sector.

For a program to be successful, it is necessary not only to plan to develop improved technology but also to transfer the technology to the farmers. Conventional wisdom suggests that when new technology is both profitable and practical, and meets the requirements of the producers, it will spread even in the absence of active extension efforts. This conclusion has been drawn mainly from experience with highly profitable cash crops and, in certain cases, from the adoption of new varieties in irrigated areas in which there are concentrations of large numbers of farms that produce the same crop. However, when a new technology has to be taught to large numbers of small farmers who grow a wide range of crops and who may not have good information sources, extension work is likely to be critical in the adoption of new techniques or varieties. Nevertheless, extension efforts will fare best when the new technology is relatively simple and appropriate to the socioeconomic and agronomic constraints faced by the farmer.

National program objectives

Before a national cassava program is formed, it is essential to clearly define the objectives and the framework in which these objectives are to be achieved. A preliminary study should be made to estimate where cassava can be produced in the country, the probable commercial yields, what inputs are required by the farmers, and what processing plants and infrastructure are needed. These estimates will give a rough idea of the costs of production, both to the farmer and to the state. From these calculations, a price can be put on cassava roots that can be used to assess the potential demand for cassava for traditional or novel uses. In order to satisfy the demand, improvement in infrastructure—such as roads to facilitate transport or investment in processing plants—may be required. The cost of these improvements should also be estimated and used in making a full-scale feasibility study that will more accurately determine cassava yields and required inputs. This information can then be used to decide whether a full-scale program should be embarked upon and, if so, to define the program objectives and the resources required to meet them.

Potential production areas

Cassava will grow in many soils and under a wide range of climatic conditions, but like most crops, it yields best in fertile

Although cassava as a species has broad adaptability, varieties must be selected individually for each agroclimatic zone. (*Source:* CIAT)

soil with little or no drought stress. Under prime agricultural conditions, farmers logically choose to grow crops that have a higher market value than cassava. On the other hand, cassava often grows and produces well in poorer agricultural areas in which few other crops can be grown without costly soil amendments or infrastructure for irrigation. In areas that are considered marginal for agricultural exploitation because the soil is acid and infertile, because the rainfall is uncertain, or because long dry periods (up to 5 months) occur, the potential for cassava production appears great because with a relatively small investment, unproductive lands can begin to contribute to the nation's total agricultural output.

It is not easy to predict the cassava yields, or the inputs required to obtain them, from marginal lands. Experimental plots can be established in a zone, but this procedure only gives an estimate of the maximum potential production with the known technology. The long-term average yield that can be expected over a large area is considerably less than that potential because as the area planted to cassava expands, soil fertility declines, the incidence of disease and pest problems tends to increase, and yields fall. In a production program for cassava alcohol established in Minas Gerais, Brazil, in the 1970s, the first small-plot yields and initial commercial yields were quite high, but as the area planted increased, yields declined drastically. Only now, as a result of research, are the yields being raised again.

There is no standard method for determining how much yields will drop as the area planted grows; nevertheless it is possible for a team of experienced agronomists, soil scientists, pathologists, and entomologists to assess the probable yield level of commercial production in a given region and the inputs required to achieve such a level. If their yield estimates are high enough to warrant further interest in the project, commercial-scale trials should be conducted to obtain more precise data on the farm yields that can be attained. This process is seldom fast enough for policy-makers, who are keen to see projects move ahead rapidly, yet such trials are essential to reduce the high risk of a failed project.

Crude estimates of the commercial cassava yield that is possible under different growing conditions and the major inputs required are presented in Table 22. The assumptions underlying the data in this table are that management is good, that the crop is kept well weeded by hand or with herbicides, that good quality planting material is used, and that integrated pest management is practiced. This table can only serve as a guideline for planning purposes: The actual yields that will be commercially obtainable in a region must be determined through commercial-scale testing.

Table 23, which is based on Table 22, gives an estimate of cassava production costs on prime and marginal land at two daily labor rates and at two different yield levels. The actual production costs will vary from locality to locality, but based on the assumptions used to make these estimates (see Appendix 1), local data on prices and land value can be used to obtain a rough estimate of the probable cost of producing cassava under different circumstances.

Table 22
Potential yields of improved cassava varieties[a] with good management under various growing conditions

Rainfall	Months to harvest	Soil fertility	Fungicides and insecticides	Planting season	Fertilizer	Yield (t/ha)
			Annual temperature range: 22–28°C			
1000+ mm well distributed	10–12	High	Stake treatment	All year	To maintain fertility	Over 35
1000+ mm 3–4 dry months	10–12	High	Stake treatment	Start or end of rainy season	To maintain fertility	30–35
1000+ mm well distributed	10–12	Moderate	Stake treatment	All year	To maintain fertility	30–35
1000+ mm 3–4 dry months	10–12	Moderate	Stake treatment	Start or end of rainy season	To maintain fertility	25–30

1000+ mm well distributed	10–12	Acid infertile	Stake treatment (+ zinc)	All year	75:150:100 NPK 0.5 t/ha lime	20–25
1000+ mm 3–4 dry months	10–12	Acid infertile	Stake treatment (+ zinc)	Start or end of rainy season	75:150:100 NPK 0.5 t/ha lime	15–20
		Annual temperature range: 20–24°C (cool winter, hot summer)				
1000+ mm mostly in summer	10–20	Moderate	Stake treatment	Spring or autumn	To maintain fertility	20–40 (depends on length of growth cycle)
		Annual temperature range: 18–22°C				
1500+ mm well distributed	14–18	Moderate	Stake treatment	All year	To maintain fertility	20–30

Source: Revised from J. H. Cock and J. K. Lynam, Cassava: Future potential and development needs, in *Proceedings: Fifth International Tropical Root Crop Symposium, Manila, Phillippines, Sept. 1979* (Los Banos: Philippine Council for Agricultural and Resources Research, 1982).

[a] Improved varieties are not yet available for all conditions.

Table 23
Production costs of cassava in relation to wage rates and yield level[a]

Land quality	Yield level (t/ha)	Wage rate (US$/day)	Production costs (US$/ton)
Marginal	15	2	36
		4	45
	25	2	22
		4	27
Prime	30	4	33
		6	37
	40	4	25
		6	28

[a] See Appendix 1 for explanation.

Potential demand

Demand for cassava must be examined in relation to the demand and price structure of other starchy crops because considerable opportunity for substitution exists. Table 24 shows the price level at which cassava would be expected to be competitive with various starchy commodities. Similar data obtained on a national or regional basis should be used to determine whether cassava can compete with other starchy staples and for what uses.

The demand for fresh cassava is more complex because of the importance of quality to the consumer. High quality, fresh cassava, particularly in the urban markets, is apparently a preferred food, and hence it must be viewed in this light rather than as a substitute for other starchy staples. The consumption of starchy staples normally decreases as income levels rise above a certain level; however there is no evidence that the consumption of fresh cassava decreases in urban areas as income levels rise. Moreover, the improved cassava storage methods that result in a high quality product with a longer shelf life may dramatically change the demand.

Table 24
Estimated price below which cassava can enter market as a substitute for competing products, in relation to yield levels and daily wage rates[a]

		Substitution price			
Cassava yield (t/ha)	Wage rate ($/day)	Milled rice ($/t)	Starch ($/t)	Sorghum or maize[b] ($/t)	Gasoline ($/l)
		Marginal land			
15	2	176	194	138	0.41
	4	221	230	166	0.46
25	2	108	138	94	0.33
	4	132	158	109	0.36
		Prime land			
30	4	162	182	128	0.39
	6	181	198	141	0.42
40	4	123	150	103	0.35
	6	137	162	113	0.36

Source: J. H. Cock and J. K. Lynam, Cassava: Future potential and development needs, in *Proceedings: Fifth International Tropical Root Crop Symposium, Manila, Philippines, Sept. 1979* (Los Banos: Philippine Council for Agricultural and Resources Research, 1982).

[a] See Appendix 1 for assumptions used to obtain data.
[b] For animal feed.

Production research

The potential of cassava is as a cheap source of energy or calories from marginal agricultural areas. It is futile to think in terms of a high-input technology based on heavy applications of fertilizers, herbicides, fungicides, and insecticides combined with irrigation, all of which are extremely costly. Rather, it is necessary to seek a production technology that can cope with the stresses that are inherent in the harsh environment in which cassava is, and will continue to be, grown.

It is easier to develop a highly productive technology for irrigated conditions with chemical control of weeds and pests and high fertilizer inputs than to design a new technology and breed varieties that are capable of producing acceptable yields under marginal conditions with little use of chemicals and without irrigation. Thus, paradoxically, the development of low-cost cassava technology requires a heavy research investment. But the return on investment in well-organized agricultural research is extremely high, and for a crop such as cassava, which science has given little attention to until recently, the payoff from research will likely be higher than it is for crops that have been more thoroughly investigated. In addition, research costs, unlike the costs for the continual purchase of inputs, can be spread over a large area over a long time period. For example, the high-yielding clone Mantiqueira, which was developed in Brazil more than 30 years ago, is still widely grown today. There is little doubt that the research investment for its development has been repaid many times over by the increased yield that has resulted from its widespread adoption by farmers. Because of the high rate of return and the long-term application of the technology that results, investment in agricultural research to resolve problems is potentially more beneficial to society as a whole than continual expenditures for chemical inputs.

International centers

Two international agricultural centers, the International Institute of Tropical Agriculture (IITA) in Nigeria and the Centro Internacional de Agricultura Tropical (CIAT) in Colombia, have large research programs devoted to cassava (see Appendix 2). These centers are developing improved varieties and agronomic practices and, to a lesser extent, methods of handling cassava roots after harvest. The IITA root and tuber improvement program is mainly concerned with developing high yielding, disease-resistant, genetic material and agronomic practices suitable for use in Africa's slash and burn culture. The IITA program emphasizes resistance to African mosaic disease, which is found in Africa and India, and biological control of mealybugs and spider mites. (In addition, IITA conducts research on farming systems that is concerned with the development of alternative, efficient, permanent crop production systems for sustained yield in the humid tropics to replace the intermittent shifting cultivation and bush-fallow systems.) CIAT is engaged in developing genetic material and technology suitable for cassava production in the Americas and Asia. CIAT also has a small research program in the development of storage and drying technologies and improved utilization of cassava products.

Researchers at these centers recognize that no single variety or production system can be appropriate for all agricultural systems in the lowland tropics where cassava is grown. IITA distributes most of its genetic material in the form of sexual seed, the progeny of which is highly variable so that national programs can select varieties that are suitable for local conditions. CIAT distributes both clonal material and sexual seed. The sexual seed is normally sent to well-established national research programs that have the capacity to handle large numbers of genetically distinct materials. Clonal material, which has already been carefully selected, is sent to the smaller national programs by both CIAT and IITA.

The centers also study the basic response of cassava to various agronomic practices. Practices suitable for certain conditions may be developed, but the objective of this research is to provide information that can be used by national or regional programs to create superior agronomic packages for their own conditions.

The research on cassava at CIAT and IITA in no way diminishes the need for strong national programs. In fact, the effectiveness of the international centers is totally dependent on the efforts of the national programs. Consequently, the international centers offer extensive training programs for personnel from national programs. In this manner, an international network of cassava researchers who freely exchange ideas, knowledge, and genetic material has been built up, and the national centers and the international centers work together to improve cassava production systems.

National research programs

Since the 1970s, there has been rising interest in cassava on the part of national agencies, and several new national research programs have been organized. One of the major difficulties in starting up these new programs has been the lack of well-trained natural and social scientists with experience in cassava to manage the programs and carry out the research. In the initial phases of program development, the national programs have made extensive use of the training courses offered by the international centers to familiarize scientists with the latest cassava technology.

Most new national programs are organized as multidisciplinary teams of scientists, in contrast to traditional research organizations in which individual scientists work within their particular discipline or sphere of interest with little concern for relating their work to that of other researchers. Scientific teams are able to resolve problems that go across traditional disciplinary boundaries. Teams of scientists are necessary for another reason: The problems in developing a new agricultural technology are so complex that one scientist working alone is unlikely to make significant progress; he or she can only provide isolated research data that cannot be easily incorporated into a technology package that can be used by the farmer.

The objectives, and hence research emphasis, of each program will vary from country to country. Nevertheless, certain research areas will most likely be common to all programs.

Varietal improvement. Since cassava production technology should be appropriate for suboptimal conditions and minimal use of purchased inputs, an ideal variety would have high yield potential, low nutrient and water requirements, and resistance to major diseases and pests. But the rate of progress in breeding any crop is generally inversely related to the number of breeding objectives. Hence, to make rapid progress, the breeding objectives should include only those factors that are not easily resolved by other means and that will eventually raise yield significantly and contribute to ease of utilization of the crop. CIAT and IITA are both developing germ plasm with superior characters, which national programs can use directly for selection or which they can incorporate in their breeding programs.

Pest management. Despite the general tolerance of cassava to diseases and pests, as production spreads and cultivation becomes more intensive, problems will increase. Recently severe losses have been caused by green spidermites, mealybugs, and cassava bacterial blight in Africa and by bacteriosis in Brazil. Adequate pest control systems should be developed, based on host-plant resistance, biological control, and cultural methods. Chemicals should be considered as a supplementary measure when other methods prove ineffective, but they should be used with care to avoid reducing the effectiveness of other control methods. Pest problems are intimately related to the local environment, so control methods must be adapted to local conditions.

For weed control, chemicals may be necessary, particularly if there is a shortage of labor. The use of manual labor and herbicides for weed control are not mutually exclusive, and integrated control methods can be developed that are appropriate for the seasonal availability of labor.

Agronomic practices. Simple improvements in agronomic practices can lead to large yield increases. Some practices, such as selection and treatment of planting material, may be fairly universal in their application while others, such as planting at the optimum time, will be very location specific. As a result, national or local agencies will need to become actively involved in these aspects of research.

A major problem of cassava growing is that soil fertility declines unless fertilizer is applied. The costs of the fertilizer can be reduced, however, by developing efficient methods of application and using cheap sources of nutrients such as rock phosphates.

Cropping systems. About 40 percent of the world's cassava is planted in association with other crops, but few scientists give mixed cropping with cassava much attention. Since the cassava crop develops slowly, short-season intercrops can be grown and harvested with little effect on final cassava yield. An early harvest of another crop enables the farmer to obtain a quick return on his investment, thus alleviating the cash-flow problems that are associated with cassava's long growth period. In addition, mixed cropping tends to decrease disease, pest, and weed problems, and the use of grain legumes may help to maintain soil nitrogen status. Mixed cropping adds to human nutrition because the consumption of grain legumes improves the protein level of cassava-based diets. There is considerable scope for improving yields from mixed cropping involving cassava. Since there has been a lack of previous research on this aspect of production, rapid progress should be possible.

Technology validation. There are many examples of technology that produces extremely good results on experiment stations but is disappointing in farmers' fields. Correct design of technology and careful attention to conducting experiments under conditions similar to those encountered by the farmer can help avoid such problems. However, at an early stage of development, technology should be validated in farmers' fields with the farmer playing a major role in growing the crop. On-farm evaluation of technology frequently exposes flaws and draws attention to the availability of inputs and resources required by the new technology. If the technology is shown to be deficient, the problems can be rectified before it is widely promoted. If required resources are not available, steps can be taken to ensure that they will be in the future. By involving local extension agents in technology evaluation trials, it is also possible to obtain information on the ease with which the new technology can be transferred.

Technology transfer. Most cassava growers routinely try new varieties on a small scale. If they like a new variety, they slowly increase the proportion of their cassava area planted to it. Yet, in the absence of a determined extension effort, the adoption of appropriate new varieties will be slow for several reasons: Cassava is frequently planted in isolated plots, which impedes communication among growers of the crop; the yield of a field of cassava is apparent only at harvest, so that neighboring farmers cannot readily notice a productive new line; the crop cycle is long, traditional propagation systems are slow, and there is little incentive for private salesmen to promote the spread of the new varieties. An extension effort to show the advantages of the new varieties, coupled with a "seed" production program, can greatly enhance the rate of adoption.

Availability of inputs. Even a cassava technology that is based on low levels of input use requires certain services and inputs. Among the most important is clean planting material of good varieties. Simple, rapid propagation systems that require little capital investment and that produce high quality planting material exist. The return on investment in propagation facilities has not been sufficiently high to attract commercial seed companies, but national agencies should be able to set up self-sustaining prop-

Facilities for rapid propagation do not need to be elaborate or expensive. (*Source:* CIAT)

agation facilities as a service to cassava growers, as has been done in Colombia.

Certain agrochemicals are necessary, too. Fertilizers are essential in infertile soils and must be applied to obtain consistently good yields on moderately fertile soils. In marginal agricultural zones, fertilizer is often expensive because of the high cost of transporting bulky commodities over poor roads. Thus, if an area with poor infrastructure is to be developed using cassava as a major crop, the construction of roads may be necessary to facilitate both the inward movement of inputs and the outward flow of products.

Small quantities of herbicides, pesticides, and fungicides are required, and the main problem in this case is not transport but availability. Chemical treatment of cassava planting material at a cost of about US$4 to US$8 a hectare gives marked improvements in yield in many instances, but farmers may not have access to the small quantities of chemicals they need. It seems reasonable to expect national agencies to establish input distribution channels (that reflect real costs) to ensure that farmers can readily obtain the agrochemical inputs they need for improved cassava production.

In recent years, the devastation caused by green spidermites in East Africa, mealybugs in Zaire, and the hornworm in parts of South America leaves no doubt about the importance of pest control, even in traditional cassava growing systems. Biological control is a powerful weapon for fighting insects; however farmers often cannot obtain the biological control agents. National agencies should provide these services or encourage commercial houses to do so.

Processing and utilization

The problems of handling cassava after harvest stem from the fact that it deteriorates extremely rapidly. All processing methods should aim at rapidly producing a more stable, less perishable product, which will reduce handling costs and risk and also facilitate marketing by allowing an accumulation of reserve stocks. Post-harvest handling technology appears to be much less location specific than production technology, and hence strong research efforts in a few locations may generate technology that can be used in many areas of the world.

Food. Recent research at the Tropical Products Institute (TPI) in London and CIAT has advanced knowledge of how to control the deterioration of fresh cassava. The two institutes have developed storage techniques that, with minor adaptations, can be used in many cassava growing areas.

A number of publicly funded institutes, including TPI and the Instituto de Investigaciones Technologicas in Bogota, Colombia, are developing improved technology for making bread that contains cassava. In fact, the technology could be used in many areas now if a sufficient amount of cheap cassava flour were available and not in competition, as so often happens, with subsidized wheat imports. Nations that have scarce resources may well be able to transfer such food technology from other areas instead of embarking on expensive research programs.

Other products. The technology for the production of chips and pellets for animal feed has been developed largely by commercial interests with some assistance from publicly funded institutions, and the situation is similar for starch and alcohol production. Generally, processing technology can be readily transferred from one region to another, so not every national research program will need to do exhaustive research on cassava processing.

Nevertheless, research effort may still be required. For example, there is a critical question concerning the concept of using cassava to produce fuel alcohol: Unless the stalks are used to fire the distillation boilers, or the whole process is made to require much less energy, the net energy gain from cassava alcohol is likely to be small. Such problems can only be resolved by extensive research, but the results should be broadly applicable to many areas.

Government policy

Expanded cassava production could contribute to greater total agricultural production, improve the nutrition of the lower income groups, reduce dependence on imported grains, and even moderate demand for imported oil by serving as an energy source. In addition, cassava offers a means to bring underutilized, marginal lands into cultivation and to create jobs in such areas, where production may be substantial. But increasing production may require certain policy changes.

Increased cassava production can only be obtained if good technology is available for efficient production of the crop. Cassava has been neglected by most government research agencies, so few tropical countries have good technology that has been proved on the farm level. A first step toward reaping the benefits of increased cassava production must be the support of research and development of low-cost production technology. This technology must permit cassava to be produced at prices low enough for the crop to be competitive with other starchy staples and with yields that are high enough to give farmers sufficient profits to make expansion of production worthwhile.

Government policies often interfere with the substitution of cassava for other starchy staples, particularly grains. As the developing world has changed from being a net exporter to being a net importer of food products, governments strive to maintain low urban food prices. This goal has often been achieved by directly or indirectly subsidizing imported or locally produced grains. Subsidies for grains, whether direct or in such forms as cheap credit or subsidized inputs, diminish the economic incentives to expand cassava production. Support prices for grains, however, do not work against cassava—they tend to support the price of cassava, too, where it can substitute for grains. A realization of the potential utility of cassava, coupled with policies that support cassava research, development, and extension and do not militate against cassava by subsidizing competing products, can open the way for increased cassava production and better utilization that will result in greater self-sufficiency in food for many developing nations.

Appendix 1: Estimation of production costs of cassava and competitiveness with other energy sources

The figures and calculations shown in this section can be used to obtain a preliminary idea of the production costs of cassava and its ability to compete in the marketplace with other starch staples (values are in U.S. dollars).

Production costs

	Production costs per hectare
Return on land and management	
Prime land	$500
Marginal land	$100
Stake treatment	$ 8
Fertilizer costs (using cheap sources of phosphorus such as rock phosphate)	
Prime land	$ 75
Marginal land	$150
Chemical weed control (optional, can be replaced by manual control)	$ 50
Labor requirements	
Manual land preparation and weeding	110 man-days
Mechanical land preparation, manual weeding	90 man-days

Mechanical land preparation, early weed control by chemical means	65 man-days
Mechanical land preparation	$100

The production costs on marginal land, with a reasonable return on land of $100/ha per year, using mechanical land preparation and chemical weed control and with a daily labor rate of $4, would be as follows:

	Production costs per hectare
Return on land and management	$100
Fertilizer	150
Stake treatment	8
Chemical weed control	50
Labor (65 days at $4/day)	260
Land preparation	100
Total	$668

It is also necessary to know what yield could be expected on a commercial basis using the inputs outlined above. Assuming the yield is 25 t/ha, then production costs on marginal land would be approximately $27/t of fresh cassava.

Comparison with other staples

If the per-ton production cost of fresh cassava, allowing for farmer profit, is known, then the price at which *fresh* cassava becomes competitive with other commodities can be estimated from the following equations:

As a substitute for rice

$$P_{\text{rice}} = 4.9\,P_{\text{cassava}}$$

As a source of industrial starch

$$P_{\text{starch}} = 4\,P_{\text{cassava}} + 50$$

As an animal feed

$$P_{\text{maize or sorghum}} = 3.125\ P_{\text{cassava}} + 25$$

As a source of alcohol

$$P_{\text{alcohol}} = (P_{\text{cassava}}/170) + 0.2$$

where *P* stands for price per ton for the cereals and fresh cassava and price per liter for alcohol.

Thus if fresh cassava is produced for \$27/t, as in the example given above, it would be competitive with sorghum or maize priced at

$$3.125 \times 27 + 25 = 109.375$$

that is, \$109.38/t, or more. Similarly, if fresh cassava is produced for \$27/t, it would be an alternative source of fuel alcohol if the latter is priced at

$$27/170 + 0.2 = 0.359$$

or \$0.36/l, or more.

The major assumptions in the equations are

- *Rice*: Fresh cassava and rice are compared, allowing for a large marketing margin for fresh cassava and also allowing for a 20 percent higher value for rice due to its higher protein content.
- *Starch*: For starch production, the figures used are derived from small-scale extraction plants, which tend to have high production costs and low extraction rates.
- *Chips*: The production cost of chips is estimated at \$25/t of chips, which is somewhat higher than the estimates of costs in Thailand. The cassava chips are assumed to have a market value of 80 percent of that of sorghum or maize for animal feed.
- *Alcohol*: Alcohol production is assumed to be large scale, 170 liters of alcohol per ton of fresh cassava, with a processing cost of \$0.20/l.

Appendix 2: Where to get technical assistance and information on cassava

To get national programs moving, funding, technical assistance, and information from outside sources can be very helpful. This appendix describes only those agencies that have a specific interest in cassava. General sources of agricultural assistance are discussed in *Agricultural assistance sources* published by the International Agricultural Development Service.

Centro Internacional de Agricultura Tropical
AA 67–13, Cali, Colombia

CIAT is one of the major sources of new genetic material and improved technology for cassava growing. Although CIAT itself is situated at an altitude of 1000 meters, most of its varietal selection and technology development is carried out in hot, lowland sites that are representative of many of the cassava growing areas of the world.

CIAT has a group of scientists working as a team to develop inproved technology, based on low-cost purchased inputs, to improve the productivity and utilization of cassava. The research program concentrates on crop physiology, entomology, pathology, plant nutrition, germ plasm collection and classification, plant breeding, agronomy, tissue culture, and various aspects of utilization of cassava. With more than 2500 clones collected throughout the Americas, CIAT has the most comprehensive germ plasm collection in existence. This collection has been screened for useful agronomic practices. Clones with specific characters can be provided to national or local programs on request. The results of the research at CIAT are published in the annual reports of the center, which are available in both English and Spanish.

Among the main products of the research and development program are new varieties and sexual seed derived from controlled crosses made to obtain specific combinations of characters. These materials can be obtained from CIAT for testing and evaluation by national agencies. Clonal material is available in the form of tissue culture, which minimizes the risk of introducing new diseases and pests.

CIAT scientists have experience in the diagnosis of production problems and also program development. These scientists are free to visit any country and assist in the resolution of problems. Because of the pressure of requests for consultation, the scientists can usually visit countries for only short periods. Requests for help from African countries are normally referred to IITA (see below).

CIAT has a large training program, which ranges from short, intensive production courses to in-service training in which students may do thesis research for higher degrees. CIAT sometimes also helps with training courses run by national agencies.

The Cassava Documentation Center brings together published information on cassava in the form of summarized and annotated bibliographies that are produced and updated on an annual basis. In addition, literature searches on specific subjects or combinations of subjects can be obtained. The documentation center also publishes a series of monographs and review articles about cassava. The services of the documentation center can be obtained for a small fee. The center is also always pleased to receive formal or informal information to be used in the *Cassava newsletter*, which is published quarterly.

International Institute of Tropical Agriculture
P.M.B. 5320, Ibadan, Nigeria

IITA emphasizes development of an improved slash and burn culture. Within this framework, the tropical root crop program works with cassava, yams, and sweet potatoes. The cassava work is oriented toward developing improved germ plasm that is resistant to African mosaic disease and cassava bacterial blight. Most material is available in the form of sexual seed for later selection and evaluation by national agencies. Research is also done on agronomic practices that are suitable for African conditions.

IITA offers consultation and training that are especially tailored to meet the needs of African national programs. Research results are published in the IITA annual report.

Tropical Products Institute
56–62 Gray's Inn Road, London WC1X 8LU, United Kingdom

TPI researches and gives technical assistance on post-harvest handling and marketing of a wide range of tropical products. It has considerable experience in cassava research, particularly in regard to fresh storage, solar drying, and the use of cassava as a wheat substitute in bakery products. The staff members spend much of their time overseas working with national program members. Requests for such assistance should normally be placed through official government channels.

Asian Institute of Technology
P.O. Box 2754, Bangkok, Thailand

AIT is both a research and a graduate training center. The scope of the work is broad, and in recent years, the institute has been directly involved in the development of improved cassava drying systems. AIT offers excellent graduate training opportunities for Asians.

Central Tuber Crops Research Institute
Trivandrum 695017, Kerala, India

CTCRI is the base for the oldest major cassava program in the world. CTCRI has developed extremely effective agronomic practices and has a breeding program for the acid-infertile soils of southern India. In addition, a great deal of attention has been paid to technology transfer and post-harvest processing. Results from CTCRI activities are published in the annual reports and in the *Journal of Root Crops*.

Commonwealth Institute of Biological Control (CIBC)
Gordon St., Curepe, Trinidad

CIBC has been active in collecting biological control agents for use with cassava. Stocks of predators and parasites are kept, and CIBC will help, on request, with the introduction and development of biological control problems.

International Development Research Center (IDRC)
P.O. Box 8500, Ottawa, Canada K1G 3H9

IDRC has been one of the main sources of funds for the development of cassava projects by both international and national centers. The agency offers technical advice on setting up projects, financial assistance to run them, and scholarships for training personnel to work in and also to manage the projects. In addition, IDRC sponsors training courses and workshops on cassava. Proceedings of the workshops have been published by IDRC.

International Tropical Root and Tuber Crop Society
c/o Dr. K. Caesar, Technical University of Berlin, Institute of Crop Science
Albrecht-Thaer-Weg 5, D-1000 Berlin 33, German Federal Republic

The International Tropical Root and Tuber Crop Society was founded by a group of scientists, mainly from the West Indies. It brings scientists together every 3 years at a symposium to discuss research findings. The symposia proceedings provide a wealth of information on cassava and other root crops.

Bibliography

Albuquerque, M. de, and Ramos-C., E. M. 1980. *A mandioca no tropico umido.* Brasilia: Editena.

This book, in Portuguese, brings together the lifetime experience of Albuquerque in working with cassava in the Amazon basin. It is extremely useful for anyone interested in cassava growing in the hot, humid, lowland tropics.

Araullo, E. V.; Nestel, B.; and Campbell, M., eds. 1974. *Cassava processing and storage: Proceedings of an interdisciplinary workshop, Pattaya, Thailand, 17–19 April 1974.* Ottawa, Canada: International Development Research Centre.

Concentrates on the chipping and drying in Asia. The comparisons of systems used by different countries are particularly useful.

Asher, C. J.; Edwards, D. G.; and Howeler, R. H. 1980. *Nutritional disorders of cassava.* Brisbane: University of Queensland, Department of Agriculture.

A well-illustrated, comprehensive booklet that describes the deficiency symptoms of nutrients essential for cassava growth. Methods for correcting the deficiencies are also given.

Bellotti, A., and van Schoonhoven, A. 1978. *Cassava pests and their control.* Cali, Colombia: Centro Internacional de Agricultura Tropical.

Describes the identification and control of cassava pests. Particular attention is given to assessing losses and integrated pest management using minimal chemical control.

Black, R. P.; Peyayopanakul, W.; and Piyapongse, S. 1978. *Thailand: Cassava pelletizing technology.* Denver: University of Denver Institute of International Programs.

Gives a detailed description of the cassava pelleting industry in Asia, including production costs and capital investment. The advantages of hard and soft chips are discussed relative to the price premiums needed to make hard chips economically viable.

Booth, R. H., and Wholey, P. W. 1978. Cassava processing in Southeast Asia. In *Cassava Harvesting and Processing Workshop held at CIAT, Cali, Colombia, 24–28 April 1978,* ed. E. J. Weber, J. H. Cock, and A. Chouinard, 7–11. Ottawa, Canada: International Development Research Centre.

This is a useful and concise summary of cassava production and processing in Asia.

Breckelbaum, T.: Bellotti, A.; and Lozano, J. C., eds. 1978. *Proceedings: Cassava Protection Workshop, Cali, Colombia, 1977.* Cali, Colombia: Centro Internacional de Agricultura Tropical.

Recent developments in crop protection and cassava storage are discussed at length. Particular attention is paid to integrated pest management and future development needs.

Breckelbaum, T.; Toro, J. C.; and Izquierdo, V. 1980. *1° Symposio Colombiano sobre Alcohol Carburante.* Cali, Colombia: Centro Internacional de Agricultura Tropical.

This symposium explored the possibilities of producing alcohol, as a gasoline substitute, from various crops, including cassava, in Colombia.

Bunting, A. H. 1979. *Science and technology for human needs, rural development, and the relief of poverty.* IADS Occasional Paper. New York: International Agricultural Development Service.

Concentrates on the gap between the scientist and the farmer and the problems faced by both in developing countries.

Burgess, Thomas. 1979. Thailand: Can it preserve its biggest money maker? *Agribusiness World* Sept./Oct., 42–47.

Describes the marketing and processing innovations behind the development of the Thai cassava pellet industry. The problems of low quality and political pressure from European farmers to restrict importation are discussed.

Buringh, P., and ven Heemst, H. D. 1977. *An estimate of world food production based on labour-oriented agriculture.* Wageningen, Netherlands: Centre for World Food Market Research.
Critically examines the uses of inputs in agriculture and concludes that fertilizer inputs are essential to feed the world even at today's population levels. The problems of equity and small farmers are clearly separated from the food production problem.

Butler, E. J.; Brown, E. E.; and Davis, L. H. 1971. *An economic analysis of the production, consumption, and marketing of cassava (tapioca).* Research Bulletin 97. Athens: University of Georgia, College of Agriculture Experiment Station.
This comprehensive review article is particularly strong on labor requirements for production.

Central Tuber Crops Research Institute. *Annual report.* Kerala, India.
Describes developments in agronomic practices, breeding, and utilization of cassava by this institute.

Centro Internacional de Agricultura Tropical. 1982. Cassava production systems program. *Annual report 1981.* Cali, Colombia.
Summarizes progress in research and development by the CIAT cassava program. This and the other annual reports, which started in 1969, cover a wide range of subjects on cassava, from biological to economic aspects.

Clark, H. E. 1978. Cereal-based diets to meet protein requirements of adult man. *World Review. Nutr. Diet.* 32:27–48.
Reviews minimal protein requirements according to the available evidence.

Cock, J. H. 1979. Cassava research. *Field Crops Research* 2:185–191.
Summarizes the major research advances in cassava production.

Cock, J. H., and Howeler, R. 1978. The ability of cassava to grow on poor soils. In *Crop tolerance to suboptimal land conditions,* ed. G. A. Jung, 145–154. Madison, Wis.: American Society of Agronomy.
Reviews cassava's ability to grow on marginal soils and pays special attention to soil fertility status.

Cock, J. H., and Lynam, J. K. 1982. Cassava: Future potential and development needs. In *Proceedings: Fifth International Tropical Root*

Crop Symposium, Manila, Philippines, Sept. 1979. Los Banos: Philippine Council for Agricultural and Resources Research.
Discusses the potential of cassava to meet shortfalls in production of calories in the developing countries and the research, development, extension, and government policy requirements necessary to allow its potential to be reached.

Cock, J. H.; MacIntyre, R.; and Graham, M., eds. 1977. *Proceedings of the Fourth Symposium of the International Society for Tropical Root Crops held at CIAT, Cali, Colombia, 1–7 August 1976.* Ottawa, Canada: International Development Research Centre.
Covers such topics as the origin and dispersal of cassava, production practices, disease and pest problems, utilization and processing, and socioeconomic aspects.

Cock, J. H.; Wholey, D.; and Lozano, J. C. 1976. *A rapid propagation system for cassava.* Cali, Colombia: Centro Internacional de Agricultura Tropical.
Describes how to construct and operate a simple rapid propagation system for cassava.

Conceicao, A. J. da. 1979. *A mandioca.* Cruz das Almas, Brazil: Universidade Federal da Bahia, Escola de Agronomia.
This textbook in Portuguese brings together much of the early research done on cassava in Brazil.

Coursey, D. G., and Haynes, P. H. 1970. Root crops and their potential as food in the tropics. *World Crops* 22(5):261–265.
Emphasizes the importance of cassava and other root crops in the nutrition of tropical peoples and analyzes the potential for higher production.

Cresswell, D. C. 1978. Cassava as a feed for pigs and poultry. *Review Trop. Agri.* 55:273–282.
A concise review of the use of cassava for pigs and poultry.

Denevan, W. M. 1971. Campa subsistence in the Gran Pajonal Eastern Peru. *Geographical Review* 61:496–518.
An interesting account of the Campa way of life and the dependence of the Campa on cassava. Their shifting culture systems are described in detail.

deVries, C. A.; Ferwerda, J. D.; and Flach, M. 1967. Choice of food crops in relation to actual and potential production in the tropics. *Neth. J. Agr. Sci.* 15:241–248.

Analyzes the potential productivity of various tropical food crops and concludes that cassava has an extremely high potential for productivity.

Diaz, R. O.; Pinstrup-Andersen, P.; and Estrada, R. D. 1975. *Costs and use of inputs in cassava production in Colombia: A brief description.* Cali, Colombia: Centro Internacional de Agricultura Tropical.

An exhaustive study of the production practices, yields, and utilization of cassava under widely different ecological and socioeconomic conditions in Colombia. The study places emphasis on labor requirements for production and factors that constrain yields on the farm level.

Doll, J. D., and Piedrahita, W. 1976. *Methods of weed control in cassava.* Cali, Colombia: Centro Internacional de Agricultura Tropical.

Describes the principles of weed control in cassava. The section on chemical control is now somewhat out-of-date.

Evenson, R. E. 1978. The organization of research to improve crops and animals in low income countries. In *Distortions of agricultural incentives,* ed. T. W. Schultz, 223–245. Bloomington: Indiana University Press.

Discusses requirements for research on crop improvement. The low research budget for tropical root crops in the developing countries is highlighted.

Ezeilo, W.N.O.; Flinn, J. C.; and Williams, L. B. 1975. *Cassava producers and cassava production in the East Central State of Nigeria.* Ibadan, Nigeria: National Accelerated Food Production Project.

A benchmark study of production practices in one state of Nigeria; concentrates on socioeconomic factors involved in the production process.

Flinn, J. C., and Lagerman, J. 1980. Evaluating technical innovations under low resource farmer conditions. *Exp. Agr.* 16:91–101.

A methodology for on-farm technology valuations. Also presents interesting data on the relationships between yield, human population density, and soil fertility.

Food and Agriculture Organization (FAO). 1980. *Food balance sheets 1975–1977 average.* Rome.

A good guide to different uses of cassava in different countries and also to the relative importance of cassava in the diet.

______. 1981. *Production yearbook.* Vol. 35. Rome.

The best available source of trends in yield productivity and production on a global basis that is readily available for most important crops.

Goering, J. 1979. *Tropical root crops and rural development.* Staff Working Paper no. 324. Washington, D.C.: World Bank.

Examines the demand for tropical root crops and concludes that there is a great potential demand for food and animal feed. Suggests that production and utilization research, as well as extension to transfer new technology to the farmers, should be emphasized.

Gomez da Silva, J.; Serra, G. E.; Moreira, J. R.; Goncalves, J. C.; and Goldemberg, J. 1978. Energy balance for ethyl alcohol production from crops. *Science* 201:903–906.

Estimates the energy balance for both the agricultural process and the industrial process. The review is somewhat pessimistic about the possibilities of using cassava as an energy source as the use of stems is entirely discounted.

Grace, M. 1971. *Cassava processing.* Agriculture Series Bulletin No. 8. Rome: FAO.

Describes cassava processing equipment and also labor and capital requirements.

Hahn, S. K.; Terry, E. R.; Leuschner, K.; Akobunda, I. O.; Okali, C. O.; and Lal, R. 1979. Cassava improvement in Africa. *Field Crops Research* 2:193–226.

Describes the major problems of and constraints on yield of cassava in Africa. A research effort to resolve these problems is described in considerable detail.

Hendershott, C. H.; Ayres, J. C.; Brannen, S. J.; Dempsey, A. H.; Lehman, P. S.; Obioba, F. C.; Rogers, D. J.; Seerley, R. W.; and Tan, H. K. 1972. *A literature review and research recommendations on cassava.* Athens: University of Georgia.

Covers all aspects of cassava with emphasis on future research. An extremely useful reference work with information from numerous difficult-to-obtain sources.

Hopper, W. D. 1976. The development of agriculture in developing countries. *Scientific American* 235:196–204.
Analyzes the need for improved technology and additional support from developed countries to enable the less-developed countries to improve agriculture.

Ingram, J. S. 1972. *Cassava processing: Commercially available machinery.* London: Tropical Products Institute.
Gives a list of commercially available processing equipment, but does not give an opinion on performance or efficiency.

International Agricultural Development Service. 1981. *Agricultural development indicators.* New York.
Provides data for comparing agricultural development progress of nations.

______. 1982. *Agricultural assistance sources.* 4th ed. Arlington, Va.
Describes the activities and interests of over 35 organizations that offer financial and technical assistance to developing countries.

International Institute of Tropical Agriculture. 1982. Tropical root crops program. *Annual report 1981.* Ibadan, Nigeria.
This and other reports since 1971 summarize the research achievements of this institute and cover a broad range of topics related to cassava.

Jones, W. O. 1959. *Manioc in Africa.* Stanford: Stanford University Press.
Onc of the classic works on cassava. Written in an easily understood style, it describes the introduction, spread, and use of cassava throughout the African continent. Although it has an African focus, it is a valuable reference book for other regions.

Kaneda, H., and Johnston, B. F. 1961. Urban food expenditure patterns in tropical Africa. *Food Research Institute Studies* 2:229–275.
Deals with the whole range of food items in the diets of tropical Africa and includes much useful data on cassava consumption and movement to urban markets.

Lancaster, P. A.; Ingram, J. S.; Lim, M. Y.; and Coursey, D. G. 1982. Traditional cassava based foods: Survey of processing techniques. *Econ. Botany* 36:12–45.
An up-to-date and comprehensive review of traditional cassava processing.

Leon, J. 1977. Origin, evolution, and early dispersal of root and tuber crops. In *Proceedings of the Fourth Symposium of the International Society for Tropical Root Crops held at CIAT, Cali, Colombia, 1–7 August, 1976,* ed. J. H. Cock, R. MacIntyre, and M. Graham, 20–36. Ottawa, Canada: International Development Research Centre.
A highly readable article that follows the domestication and spread of root crops throughout the tropics.

Lozano, J. C.; Bellotti, A.; Reyes, J. A.; Howeler, R.; Leihner, D.; and Doll, J. 1981. *Field problems in cassava.* 2nd ed. Cali, Colombia: Centro Internacional de Agricultura Tropical.
Describes how to identify common field problems related to diseases, pests, herbicides, and nutrient deficiency. The use of color photographs makes this book easy to use.

Lozano, J. C., and Booth, R. H. 1974. Diseases of cassava. *PANS* 20:30–54.
Covers the major diseases of cassava with special reference to control measures and possible losses.

Lozano, J. C.; Cock, J. H.; and Castano, J. 1978. New developments in cassava storage. *Proceedings: Cassava Protection Workshop, Cali, Colombia, 1977,* ed. T. Brekelbaum, A. Bellotti, and J. C. Lozano, 135–141. Cali, Colombia: Centro Internacional de Agricultura Tropical.
Information on methods for storing fresh cassava.

Lozano, J. C.; Toro, J. C.; Castro, A.; and Bellotti, A. 1977. *Production of cassava planting material.* Cali, Colombia: Centro Internacional de Agricultura Tropical.
Describes how rather simple methods can be used to obtain good quality planting material, one of the most important aspects of the new technology.

Luzuriaga, H. 1976. *Descripcion agro-economica del proceso del cultivo de yuca en el Ecuador.* Publicacion Miscelanea no. 33. Quito, Ecuador:

Instituto Nacional de Investigaciones Agropecuarias, Departamento de Economia Agricola.

A detailed account of cassava production practices in Ecuador with special attention given to yield and labor requirements.

Lynam, J. K. 1978. Options for Latin American countries in the development of integrated cassava production programs. In *The adaptation of traditional agriculture: Socioeconomic problems of urbanization,* ed. E. K. Fisk, 213–256. Monograph no. 11. Canberra: Australian National University Development Studies Centre.

The status of cassava production and utilization in Latin America and the future potential is delineated.

Montaldo, A. 1979. *La yuca o mandioca.* San Jose, Costa Rica: Instituto Interamericano de Ciencias Agricolas.

This comprehensive work on cassava takes a very Latin American stance. It is a good text for university courses.

Nestel, B., and Cock, J. H. 1976. *Cassava: The development of an international research network.* Ottawa, Canada: International Development Research Centre.

Describes the development of an integrated worldwide research effort on cassava in the early 1970s. Emphasizes projects supported by the IDRC.

Nestel, B., and Graham, M., eds. 1977. *Workshop on cassava as an animal feed, University of Guelph, 1977.* Ottawa, Canada: International Development Research Centre.

The theme of cassava as animal feed is exhaustively reviewed with special attention to formulation of rations. In the final section, areas that need further research and development are highlighted.

Nestel, B., and MacIntyre, R., eds. 1973. *Chronic cassava toxicity: Proceedings of an interdisciplinary workshop held at London, England, 29–30 January 1973.* Ottawa, Canada: International Development Research Centre.

Discusses chronic cyanide toxicity related to high cassava consumption. A complete and unbiased treatment of the subject.

Normanha, E. S., and Pereira, A. S. 1950. Aspectos agronomicos da cultura da mandioca. *Bragantia* 10(7):179–202.

Normanha, one of the pioneers of agronomic research on cassava, brings together much of his experience in developing improved agronomic practices for cassava in the state of Sao Paulo, Brazil.

Nye, P. H., and Greenland, D. J. 1960. *The soil under shifting cultivation.* Technical Communication no. 51. Farnham Royal, Eng.: Commonwealth Agricultural Bureaux.

Slash and burn culture is of greatest importance in Africa, and this scholarly work carefully analyzes this cultural system. The work is of particular interest to those people who are concerned about problems of soil conservation under primitive agricultural conditions.

Oke, O. L. 1979. Some aspects of the role of cyanogenic glycosides in nutrition. *World Rev. Nutr. Diet.* 33:70–103.

The role of cyanogenic glycosides (which are responsible for the cyanide in cassava) is looked at in depth, and several possible beneficial effects are discussed. The article shows some bias toward the favorable effects, which are not as clear-cut as is suggested.

Onwueme, I. C. 1978. Cassava. In Onwueme, *Tropical Root Crops,* 109–163. Chichester, Eng.: John Wiley and Sons.

Contains a small section on cassava with particular reference to African conditions.

Phillips, T. P. 1974. *Cassava utilization and potential markets.* Ottawa, Canada: International Development Research Centre.

The potential use of cassava for animal feed, particularly as an export crop, is exhaustively studied. The potential demand in other areas is discussed, but not studied in depth.

Poleman, T. T. 1961. The food economies of urban middle Africa: The case of Ghana. *Food Research Institute Studies* 2:121–175.

A detailed study of cassava use in Ghana; of particular interest for the data on the movement of cassava to the urban centers and the data on consumption patterns at different income levels.

Renvoise, B. S. 1973. The area of origin of *Manihot esculenta* as a crop plant: A review of the evidence. *Economic Botany* 26:352–360.

Examines the domestication of cassava and its subsequent spread throughout the Americas.

Schultz, T. W. 1979. *The economics of research and agricultural productivity.* IADS Occasional Paper. New York: International Agricultural Development Service.

A concise and direct analysis of the agricultural needs of developing countries; emphasizes the need for investment in research.

Subrahmanyan, V; Rama-Rao, G.; Murthy, H.B.N.; and Swaminathan, M. 1958. The effect of replacement of rice in a poor vegetarian diet by tapioca macaroni on the general health and nutritional status of children. *British Journal of Nutrition* 12:353–358.

Discounts many of the arguments against cassava as an extremely poor food source; in fact, the children grew as well on the tapioca diet as on the rice diet.

Terra, G.J.A. 1964. The significance of leaf vegetables, especially of cassava in tropical nutrition. *Tropical and Geographical Medicine* 2:97–108.

A description of the use of cassava leaves and their nutritional value.

Terry, E. R.; Oduro, K. A.; and Caveness, F., eds. 1981. *Tropical Root Crops: Research Strategies for the 1980s. Proceedings of the First Triennial Root Crops Symposium of the International Society for Tropical Root Crops—Africa Branch, 8–12 September 1980, Ibadan, Nigeria.* Ottawa, Canada: International Development Research Centre.

This symposium brought together a wealth of information on cassava in Africa. The emphasis was on breeding, disease and pest control, and agronomy. The paper by F. W. Nweke on consumption patterns of root crops in tropical Africa is of special interest.

United Nations. 1975. *Poverty, unemployment, and development policy: A case study of selected issues with reference to Kerala.* ST/ESA/29. New York.

A very complete study of the socioeconomic parameters of an area that is highly dependent on cassava and cassava products. The sections on diet and health are of particular value to cassava workers.

United Nations Conference on Trade and Development/General Agreement on Tariffs and Trade (UNCTAD/GATT). 1977. *Cassava: Export potential and market requirements.* Geneva.

Contains data on world trade in cassava with special detail on the European animal feed market. The specifications of quality for different markets are carefully defined.

Weber, E. J.; Cock, J. H.; and Chouinard, A., eds. 1978. *Cassava harvesting and processing: Proceedings of a workshop held at CIAT, Cali, Colombia, 24–28 April 1978.* Ottawa, Canada: International Development Research Centre.

Covers a wide range of topics with emphasis on mechanized harvesting, chipping, and drying.

Weber, E. J.; Nestel, B.; and Campbell, M., eds. 1979. *Intercropping with cassava: Proceedings of an international workshop held at Trivandrum, India, 27 Nov.–1 Dec. 1978.* Ottawa, Canada: International Development Research Centre.

Reviews knowledge on cassava intercropping. Great potential appears to exist for intercropping with grain legumes.

Weber, E. J.; Toro, J. C.; and Graham, M., eds. 1980. *Cassava cultural practices: Proceedings of a workshop held in Salvador, Bahia, Brazil, 18–21 March 1980.* Ottawa, Canada: International Research Development Centre.

Covers cassava agronomy. A useful review of agronomic practices in different parts of the world.

Wortman, S., and Cummings, R. W., Jr. 1978. *To feed this world.* Baltimore: Johns Hopkins University Press.

The development of national agricultural programs and several examples of successful cases are described. The book is strongly focused on development of improved cereal production, nevertheless it details many of the factors that are important in increasing the production of all foods.

Index

About the Book and Author

Cassava: New Potential for a Neglected Crop
James H. Cock

Like other root crops cultivated on small farms in the tropics, cassava was neglected for some time by policymakers and scientists. In the last decade, however, considerable attention has been given to cassava as a food, as an animal feed, and as a fuel source.

In this book, James Cock, leader of the cassava program at the International Center for Tropical Agriculture (CIAT) in Colombia, brings together the latest information on improved strains, modern production systems, better processing methods, innovations in storage and marketing, and the prospects for using cassava to produce fuel alcohol. He also explores the cassava production programs of several developing countries and offers suggestions for creating an effective national cassava program. The book will be useful to both policymakers and researchers.

Dr. James H. Cock is coordinator of the cassava program at the International Center for Tropical Agriculture, Cali, Colombia.

For Product Safety Concerns and Information please contact our EU representative GPSR@taylorandfrancis.com Taylor & Francis Verlag GmbH, Kaufingerstraße 24, 80331 München, Germany

Batch number: 10725875

Printed by Printforce, the Netherlands